Coal Mining Communities and Gentrification in Japan

Tai Wei Lim • Naoko Shimazaki
Yoshihisa Godo • Yiru Lim

Coal Mining Communities and Gentrification in Japan

palgrave
macmillan

Tai Wei Lim
Singapore University of Social Sciences
Singapore, Singapore

Naoko Shimazaki
Waseda University
Tokyo, Japan

Yoshihisa Godo
Meiji Gakuin University
Tokyo, Japan

Yiru Lim
Singapore University of Social Sciences
Singapore, Singapore

ISBN 978-981-13-7219-3 ISBN 978-981-13-7220-9 (eBook)
https://doi.org/10.1007/978-981-13-7220-9

This Palgrave Macmillan imprint is published by the registered company Springer Nature Singapore Pte Ltd.
The registered company address is: 152 Beach Road, #21-01/04 Gateway East, Singapore 189721, Singapore

Foreword

If the Anthropocene is the new world of humanity's making, then there has hardly been a more important, and more controversial, actor in making it than coal. The mass production of coal provided the power for the European, and hence global, Industrial Revolutions during the nineteenth and twentieth centuries. In Britain, coal helped industry from a total of 170,000 horsepower in 1800 to 2.2 million in 1870 and 10.5 million by 1907.[1] In the developing world today, coal is still a major source of power and electricity.

However, while coal may have been a symbol of progress and modernity in the nineteenth century, it is now more likely to reflect the dark sides of the Industrial Revolution—dirty, dangerous, and exploitative. The grey skies and poisoned waters of Linfen city in Shanxi province, in China's northern coal mining region, today symbolize for many people what is means to produce and consume coal. In 2010, Linfen was rated by the World Bank as the world's most polluted city.[2] Climate change has now made the problem of burning coal more urgent than ever. It has been estimated that coal contributes 46% of CO_2 emissions worldwide and nearly three-quarters of greenhouse gases.[3]

[1] Mitchell, Timothy. 2011. *Carbon Democracy: Political Power in the Age of Oil*. Verso: London.

[2] Feinberg, David, Chinese coal-mining city is world's most polluted, dated November 5, 2018, *CNN World – International Edition*. Retrieved on October 30, 2018, from: http://edition.cnn.com/2010/WORLD/asiapcf/06/02/toxic.linfen.vbs/index.html

[3] Endcoal, Climate Change, undated, Retrieved on October 30, 2018, from: https://endcoal.org/climate-change/

However, the availability of cleaner—if not cheaper—fuels in oil and gas, rapid development of renewable energy, and the social and political mobilization around these issues have led to massive divestment in coal worldwide. For many, the Age of Peak Coal is past. In the UK, peak coal consumption was in the 1950s, while both production and consumption have been declining ever since.[4] In 2016, they reached 1800 levels, the same as what they were at the beginning of the coal boom. In the US, even with a bullish president trying to prop up a dying industry, coal demand is expected to grow by only 1%.[5] Meanwhile, France and several other European countries have pledged to be coal-free by 2021, while more than two dozen have joined the alliance for Organisation for Economic Co-operation and Development (OECD) countries to phase out coal by 2030 and developing countries by 2050.[6]

From a more global perspective, however, coal is perhaps heading not so much towards oblivion as towards a new plateau. Coal continues to supply one-third of all energy worldwide.[7] It is especially important as a source of energy in Asia. China is still the world's largest producer of coal at 3.5 billion MT per year; Indonesia is the second largest exporter of coal, and Japan its largest importer. China, Japan, South Korea, and India account for $22 billion of the $24 billion that G20 countries have planned for new coal projects, most of which will be in India and Southeast Asia.[8] Infrastructure spending to improve connectivity and economic modernization in the region, such as with China's One Belt, One Road initiative, will likely further stimulate demand for coal. Asia-Pacific's production goal is more than two times that of the rest of the world, while its consumption is nearly three times.[9]

[4] Vaughan, Adam, Coal: global demand for coal falls in 2016 for second year in a row, dated June 13, 2017, *The Guardian*. Retrieved on October 30, 2018, from: https://www.theguardian.com/environment/2017/jun/13/coal-global-demand-falls-2016-second-year-in-row-fossil-fuel

[5] Brooks, Ed and Joanna Kao, The future of coal in seven charts, dated September 19, 2017, *Financial Times*. Retrieved on October 30, 2018, from: https://www.ft.com/content/82f9087c-9cdd-11e7-8cd4-932067fbf946

[6] Bradley, Sian, The future of coal, August and September, 2018, Chatham House. Retrieved on October 30, 2018, from: https://www.chathamhouse.org/publications/twt/future-coal

[7] International Energy Agency (IEA), Coal, undated. Retrieved on October 30, 2018, from: https://www.iea.org/topics/coal/

[8] Sian, The future of coal, 2018.

[9] BP Global, Coal production, undated. Retrieved on October 30, 2018, from: https://www.bp.com/en/global/corporate/energy-economics/statistical-review-of-world-energy/coal/coal.html

Even as coal production and consumption have been declining in China, as part of Premier Li Keqiang's promise to "make skies blue again,"[10] it still supplies one-half to two-thirds of 55% of China's energy demand. In India, coal-fired power generation is expected to increase by 4% a year.

One these trends tell us is that, despite global divestment from coal, coal's history is still far from over, and its biggest chapter may be still yet to be written. Understanding how coal continues to supply energy for the world's most populous region of 4.4 billion people will be critical for imagining a sustainable future.

For these reasons, it is an honour and a pleasure to be able to recommend *Coal Mining Communities and Gentrification in Japan,* edited by my good friend and colleague Lim Tai Wei. As both editor and contributor of several chapters, Lim Tai Wei has demonstrated once again tremendous breadth as a scholar—who, in addition to published work on coal and energy resources in Northeast Asia, has also written about students' activism in Hong Kong, robotics in Japan, cultural heritage in Singapore, the Association of Southeast Asian Nations (ASEAN) diplomacy, and much more—and a regional expertise.

Bringing together scholars from Japan and Singapore, this volume offers cross-national and cross-disciplinary perspectives on this par-excellence transboundary and multidimensional topic: the Japanese coal industry. The volume examines this industry from its dawn in the nineteenth century under the Meiji Restoration—though the practice of coal mining has of course been around for much longer—to what today certainly looks like the industry's twilight, though, as this book shows, its socio-environmental legacies and some limited use will remain a part of the Japanese landscape and collective memory for many years to come. Its combination of disciplinary expertise from international relations, political science, history, cultural studies, and literature helps us to see coal not simply as an economic resource but also as an object of deep social and cultural importance.

What is perhaps most special about this volume, however, is that it offers a cradle to grave (or almost) analysis not simply of a resource product (i.e., coal) but rather of the entire industry that grew up around it. From examining how coal production was an important vector for initially opening Japan's ports to the West, to coal's vital contribution to Japan's

[10] IEA, Coal, undated.

ill-considered military conquest during World War II, to the conversion of old mining pits into urban parks and new cultural heritage sites, the authors in this volume demonstrate not only how human history shapes resource production but also how the specific resource (i.e., coal) and its production also shaped Japanese history in particular ways. In this light, the Japanese case study can be of interest to other coal-producing regions for offering not only insights into our past but also perhaps glimpses into our future.

In doing so, these authors also remind us of the importance of local history when trying to understand deeply human decisions around resource use and management. While global trends and conditions may be important factors, they are rarely on their own sufficient to explain how, when, and why a coal industry expands or contracts in any particular place, and much less what those experiences mean for the people inhabiting and making a living in those places. In this book, we learn a lot not only about coal but also about Japan and the surrounding region.

Singapore
November 2018

Jason Morris-Jung

Contents

List of Figures

List of Tables

CHAPTER 1

Introduction: Post-Mining Communities—A Multidisciplinary Analysis of Post-Closure Coal Mining Towns in Japan and Their Gentrification History

Introduction

This book project arose from the collaboration between a sociologist who worked on the coal mining industry in Japan, a developmental economist, and an area studies specialist on Japan. These scholars combined their expertise to contribute to a multidisciplinary volume that encompasses a chronological glimpse of Japanese coal mines in three phases. Having three phases provides a neat organization of contents and materials derived from fieldwork as well as from secondary resources' interpretive work. Chronically, the volume also weaves a coherent narrative of coal industrial development through different ages experiencing different environmental conditions and externalities, so comparisons and contrasts can be highlighted temporally when the materials are segmented chronologically.

The first phase consists of the transitional period from late pre-modernity to modernity, written by a veteran economist Yoshihisa Godo. It is a historical survey that stretches from the early modern period to the early post-war decades. The second chronological phase written by sociologist Naoko Shimazaki focuses on the post-war era from where chapter contributor Godo left off. This period may be considered the "golden era" of post-war coal mining history during Japan's transition from coal to oil energy. Rising wages, labour strikes, mine depletion, societal rejection

T. W. Lim et al., *Coal Mining Communities and Gentrification in Japan*, https://doi.org/10.1007/978-981-13-7220-9_1

of coal pollution, and dangerous working conditions were the impetuses for the beginning of the slow decline of the coal industry that lasted right up to the early twenty-first century. The third phase covered in this writing is situated in the contemporary era, where Tai Wei Lim (an area studies specialist focusing on Northeast Asia, particularly Japan and China) analyses current uses of former coal mining areas for gentrification, tourism, retail activities, and heritage preservation. He utilized a combination of observation studies, interpretation of secondary textual sources, fieldwork based on visual analysis, and unstructured conversations with residents and stakeholders of those mines. For a more macro perspective complementing the study on Japan's domestic coal industry in the previous chapter, Tai Wei Lim also has a subsequent chapter on the contemporary regional developments of the coal mining industry in Northeast Asia. Finally, Tai Wei Lim wraps up the volume with a study of the cultural aspects of coal mine development in Japan through an analysis of aesthetics, gender perspectives, literary arts, and culture.

Chapterization and Introduction of Contents

The following section introduces the volume chapters with some details. Godo's historical chapter lays the foundation for understanding the early development of the coal mining industry in this volume, particularly in terms of its association with agriculture and other natural resources or what Godo calls "primary industries." His chapter explains how the early modern coal mining industry got access to a large pool of labour generated by excess manpower from a declining agricultural sector which faced oversupply. Godo's chapter also informs readers about the growing demand for coal fuel as Japan entered a heavy industrialization phase in its economic development. This phase was interrupted by the Pacific War but quickly resumed its post-war trajectory. Godo's extensive use of Japanese-language sources provides a value-added Japanese perspective of the country's industrialization process and the role of coal mining industries. In narrating the emergence of Japan's coal industry, Godo has also highlighted the transition of agricultural activities to industrialization, and the chapter is also a micro-level account of the transformation of autonomously created laissez-faire work schedules ("peasant time") to highly disciplined rationalized mass production (later epitomized by Taylorism).

The next stage of post-war evolution in the coal mining industry (including its decline) is detailed by Shimazaki. Shimazaki's chapter is centred on the Mitsui Miike Coal Mine. There are at least three points of significance in studying this coal mine. First, it was the largest coal mine in Japan, and it produced the highest quality coal. Therefore, the government also tried to make the coal mine the most sustainable in terms of extraction. Because of its size, the Miike Coal Mine was economically important and politically influential. Because of its political significance, labour relations and coal strikes in Miike also garnered the most attention, particularly from the media. In 1963, the worst industrial incident involving a coal mine occurred in Miike. All these features in Miike attracted academic research interest.

Second, Miike's closure in 1997 was particularly significant. While Kushiro and Ikeshima coal mines were chronologically the last mines to close down in Japan, Miike's closure was on a far greater scale. The third significance of Shimazaki's chapter is its contribution to existing publications in this field, as there is no English-language academic writing at the moment that studies Miike's closure in detail and embedded within a conceptual/theoretical framework. Shimazaki's hypothesis is that, from a comparative perspective, there are similarities to the industrial layoffs found in other heavy industries involving blue-collared industrial workers. These common features amongst heavy industries will be examined in Shimazaki's chapter.

Lim's chapter utilizes observation studies to go on-site to observe how gentrified coal mining spaces are developed in Hokkaido's coal mining regions of historical importance. Chronologically, Lim's story takes place after the mine closures studied by Shimazaki's chapter. The bulk of the Lim's fieldwork in West Hokkaido focuses on Yubari and Mikasa. Unused coal mines were mothballed and made defunct. To make these former coal mines environmentally green and friendly, a number of stakeholders are coming up with ways to make their environments conducive for tourism and local community livelihoods, ensuring a clean environment and sustainable development. The bulk of the author's fieldwork in East Hokkaido is focused on Kushiro. The Kushiro Coal Mine operated by the KCM firm was the first Japanese coal mine facility that Lim visited which is still operational. This was the smaller-scale contemporary version of the coal mining company with limited coal production output after the coal mine closed. Therefore, it was perhaps the most important case study since Lim was able to visualize workers at work in the mine and its daily operations.

Methodology

The volume's methodological approaches can be divided into three major directions. First, the volume's authors used observation studies to go on-site to observe how gentrified coal mining spaces are used. In the process, the authors had conversations with individuals who were affected by coal mine closure or residents of these towns who were displaced and/or economically affected by the decline of this industry. In the museums, the authors' observations also recorded the artefacts on display using digital images to curate memories of the past and serve as educational tools for community residents. The authors visited some of these mines to experience life as a coal miner in order to adopt a standpoint view and apply experiential learning related to working in the coal mine. Second, the authors relied on collection of textual documents (both primary and secondary) to reconstruct a clearer picture of the former coal mining towns before and after closure. These textual and visual documents were useful to corroborate our on-site observations. They provided some empirical and quantitative data useful for the volume. Secondary resources were also useful for interpretive work. Blueprints and town plans also provided a clearer spatial layout of the former coal mining facilities in the places that were visited by the authors.

Third, the authors also selectively carried out conversations with users of gentrified green spaces, farmers in former coal mining towns, and local museum curators to understand their opinions of post-closure changes. These interviews are semi-structured, loose, and qualitative rather than quantitative and designed with open questions that give interviewees ample space to articulate their views. Some of the sites visited by the authors for this volume included Bibai, Kushiro, Asahikawa or Yubari Coal Mining Museums, and so on. The authors walked on cycling tracks and a green rail corridor built on abandoned railway tracks in Kushiro, and so on. In order to have a standpoint perspective of daily lives in the former coal mine towns, the authors stayed in accommodations near agricultural areas in a former coal mining town in Hokkaido. The authors also visited the Yubari film festival site and farming area and scoured local book shops, archives, and libraries in former coal mining towns in Hokkaido, like Bibai, Kushiro, Asahikawa, or/and Yubari.

Literature Review

Existing writings on the mentioned mines within the context of modernization, mine closures, and gentrification are mostly found in the form of policy papers instead of an edited volume. Policy reports and academic studies that date back to the 1960s were also reviewed in order to understand the subject matter from a temporal standpoint perspective drawn from that historical period; for example, Benjamin Martin's article detailed the social impact of an early mine closure from the perspective of the early 1960s. Another classic article "Some Problems of Unemployed Laborers in Coal Mining Industry" by Naoki Kusuhara details the first post-war attempt to address post-closure issues such as re-employment of retrenched workers. It is important to note that these are short articles rather than a consolidated coherent edited volume structured around a central theme. The most important monographic work in the English language is Suzanne Cutler's *Managing Decline: Japan's Coal Industry Restructuring and Community Response*, mainly written from a sociological/anthropological point of view and with information dated back to the 1999. This writing's value-add to Cutler's volume is bringing the materials up to date and also adopting a multidisciplinary perspective.

In Tai Wei Lim's previous publication (*Energy Transitions in Japan and China: Mine Closures, Rail Developments, and Energy Narratives*), Lim studied the research questions of why coal mines declined in importance in the post-war years. In that publication, Lim focused mainly on mines found in southern Japan, particularly in Kyushu like Omuta, Chikuhō, and other mines that have since closed when Japan transitioned to the use of oil starting from the 1960s. In the course of his previous research, he came across information related to the gentrification of these former mines and discovered how they have been transformed into environmentally friendly towns with smaller populations. The former mine sites have also been converted into playgrounds, skating parks, museums, and other educational facilities.

Up till that point of time, publications on the gentrification of former coal mines were limited to Suzanne Cutler's publication *Managing Decline: Japan's Coal Industry Restructuring and Community Response* as the major seminal work in this area. This volume is a continuation of Lim's previous work on former coal mines in Japan. There are some distinguishing features of this volume from previous works. In Lim's previous volume, he mainly worked on coal mines in southern Japan's island of Kyushu. In this grant, he carried out fieldwork with his Japanese colleagues in the sites of northern coal mines in Hokkaido like Kushiro, Yubari, and Bibai. Almost all existing writings of recent vintage focused mainly on physical assets when considering gentrification. They did not look into intangible cultural artefacts and textual or graphical materials when considering the gentrification process.

Contribution to Existing Literature

The volume value-adds to existing writings through the use of a multidisciplinary approach in studying the complexities of gentrification processes. The contributors' expertise can better indicate how the gentrification process takes on tangible and intangible dimensions with concrete policies directly affecting the lives of workers and their communities' livelihoods as well as intangible recollection of memories for educational purposes and to develop a local community identity. The careful archiving of memories helps to bring the town's coal industrial decline to a final closure and transition the town to a new developmental phase while preserving the character and historical consciousness of its denizens. Sociological analyses combined with textual interpretations of retrenched coal miners' resettlement make this project rather unique and informative for comprehensive scholarly study as compared to a single-disciplinary approach.

Unlike other existing literatures, this writing includes viewpoints from a contemporary lens. It adopts contemporary ideas and applications of energy use, including the use of renewable energies in former coal mining areas to power their infrastructure and industries. In the same way, existing works on Japanese coal mine closure do not cover the aftermath of the closure. Lim examines how the former coal mining town in Hokkaido started transitioning to cleaner sources of fuels and explains how the local communities became conscious of keeping the environment clean so that service and agricultural industries that require clean and pristine environments to attract tourism and grow high-quality farm products can emerge in former coal mining areas. In this aspect, Yubari and its high-quality melon cultivation is a good example. In terms of sources, Japan-based scholars Godo and Shimazaki and their extensive use of Japanese-language sources provide a standpoint view to the narrative of Japan's industrial history from the beginning of modernization (Meiji Restoration) to post-war restructuring resulting in the closure of coal mines, bringing an end to one of the most important heavy industries in Japan. Because Japan is the first modernizing economy in East Asia, its story carries significance and is important to the narrative of early modernization in East Asia regionally.

Lim also examines how a strong sense of nostalgia for lost green spaces in the form of abandoned railway tracks of Kushiro, Bibai, and Asahikawa in Hokkaido have seen these spaces reused as cycling tracks for outdoor enthusiasts or designated green corridors for walkers, hikers, and trekkers. There is also a cultural perspective to the gentrification of former coal mining areas. The volume studies the artistic transformation and gentrification process of former coal mining industrial spaces in Bibai into cutting-edge post-modernist sculpture parks like the Arte Piazza Bibai Kan Yasuda Sculpture Park, which features the works of the internationally acclaimed sculptor Kan Yasuda. Such renewal and gentrified use of spaces necessitates the need for a more in-depth understanding of usage of former coal mining spaces, green environmental policies, and tapping on renewable energy options to clean up the towns and transition these spaces to service tourism and/or high value-added agricultural economy. While the reasons behind coal mine closures and the re-employment measures to cope with its impacts are studied comprehensively in Shimazaki's works, analysing the aftermath of the closure and restoration of the community in those areas are de-privileged in current discussions/existing volumes. This writing addresses this area.

Lim's coal mine gentrification chapter is followed by a broader regional view of the coal industry in Northeast Asia. This chapter situates the discussion of the first three chapters in the context of regional developments in the coal industry. Taken as a whole, it is possible to see how coal remains an important energy resource to Northeast Asia and its three major economies of China, Japan, and South Korea. It also looks at how the regional coal trade has evolved from mainly energy imports from coal-rich regions to industrializing Northeast Asia to major Northeast Asian countries becoming net investors in the regional East Asian coal industry, including Indonesia. These investments are now moving from coal trade to higher value-added economic activities like the construction of power stations.

The regional perspective in Tai Wei Lim's final chapter is then followed by the last two chapters (before conclusion) in this volume by Tai Wei Lim and Yiru Lim. While the previous four chapters analyse the history of Northeast Asian coal development chronologically and deal with empirical case studies, Tai Wei Lim's chapter provides the intangible aspects of the industry, studying the gender perspectives, literary (written text, folk songs, film festivals, and women's roles) subtexts, and artistic expressions of coal mining stakeholders/workers/family members throughout its industrial history. It adds the soft cultural perspective to the political, economic, sociological, and historical perspectives in the previous chapters and highlights the human story in an otherwise quantitative interpretation of industrial development. Yiru Lim's chapter continues with the gender theme, examining the world of the female miner in Japan. The aim of this chapter, then, is to utilize a gendered lens to understand the socio-cultural and historical contexts and working conditions under which Japanese women miners laboured, and, consequently, to analyse points of resistance, if any, that women displayed within this environment. It is befitting as the last chapter before the conclusion, as it wraps up the story of the success, trials, and tribulations of the industry's history. Before substantial discussions on contemporary issues of coal mine closures and gentrification, the next chapter after this introduction will first turn to the historical foundations of the coal mining industry in Japan.

CHAPTER 2

Comparison of the Impacts of Japan's Industrialization on the Coal Mining and Agriculture and Forestry Industries

This chapter examines how Japan's coal mining industry changed along with the industrialization of the Japanese economy. It adopts a comparative lens for this purpose. In doing so, this chapter compares the historical changes in the coal mining industry with those in agriculture and forestry, which also belong to the primary sector. There are similarities in the historical performance of these three industries. From the proto-industrialization period to the light industrialization period, the three industries experienced relatively good performance. However, when Japan faced hardships during the pre-war heavy industrialization period, the government supported the secondary sector by suppressing the prices of primary sector commodities, which resulted in serious rural poverty. In contrast, during the post-war heavy industrialization period, the government provided enormous subsidies to the three industries. This income redistribution policy contributed to post-war Japan's high economic growth and its prerequisite of social stability by preventing social unrests. This chapter shows that, in order to gain a comprehensive understanding of the economic history of the coal mining industry, comparative studies with other industries in the primary sector are useful.

T. W. Lim et al., *Coal Mining Communities and Gentrification in Japan*, https://doi.org/10.1007/978-981-13-7220-9_2

Introduction

The coal mining industry belongs to the primary sector. According to the famous hypothesis of the Colin-Klerk model, once modern economic growth begins, major resources, such as labour force and capital, move from the primary sector to the secondary sector. This process is popularly known as industrialization. Industrialization induces many structural changes in the primary sector. This chapter examines how Japan's coal mining industry changed because of industrialization.[1] In doing so, this chapter compares the historical changes of the Japanese coal mining industry with those found in the agriculture and forestry industries, which also belong to the primary sector. This comparison will be useful in identifying the commonalities within the primary sector and the unique characteristics of the coal mining industry. In spite of its usefulness, such an approach has not been sufficiently implemented in the literature so far. This may be because researchers from different fields do not interact actively in Japan's academic society. The author, who is familiar with agricultural economics, aims to fill this gap.

Comparisons Between the Coal Mining and Agriculture and Forestry Industries

The primary sector includes production activities that directly interact with natural resources. In terms of gross domestic product (GDP) and employment, agriculture is the biggest industry in Japan's primary sector. In addition to agriculture, this study selects forestry for comparison with the coal mining industry because forestry is similar to coal mining in the sense that both wood and coal are inedible products (although mushrooms and wild vegetables are also included in forestry, they are minor

[1] The Colin-Klerk model further posits that, after an economy completes the industrialization stage, it will move to the post-industrialization stage, wherein the economic gravity moves from the secondary sector to the third. In the post-industrialization stage, the percentages of GDP and labour force of the primary sector kept declining. Japan's post-industrialization period started around the 1980s. As discussed in section "Protection for the Primary Sector During Post-war Japan's Heavy Industrialization", at that time, the coal mining industry had already shrunk to a negligible level in the Japanese economy. Thus, this chapter does not cover Japan's post-industrialization period.

products among the total value of forestry products).[2] In addition, timber is in competition with new types of materials such as steel and plastics, which are developed based on modern technologies; similarly, coal is in competition with oil and natural gas.[3] There are differences between the coal mining and forestry industries as well. First, coal mining is a non-renewable resource, whereas forestry is renewable. Second, coal is only used as a fuel, whereas wood is used not only for fuel but also as material for the construction and manufacture of equipment/tools and paper. Third, forestry is occasionally affected by natural disasters, whereas coal mining is affected by manmade, not natural, disasters.

The Primary Sector in Pre-modern Japan

Before the arrival of warships flying Western flags in the 1850s, Japan, governed by the Tokugawa Shogunate founded in 1603, was under a strict policy of seclusion (*sakoku*). The society was in a pre-modern and feudal economy. In the pre-modern economy, farmers collected both coal and wood for use in their daily activities. In that sense, before Japan's modernization started around the Meiji Restoration of 1868, neither coal mining nor forestry had become industries yet. However, there were early signs of the formation of industries. For example, the salt industry, which grew significantly in the late Tokugawa Shogunate period, had increased the demand for wood and coal as fuel. This resulted in groups of workers spending most of their work hours collecting coal or wood.

The Tokugawa Shogunate implemented a strong regulation system for farmers, called *Ryoshu Sei*. The lord of a feudal clan, called *Ryoshu,* was in charge of keeping the social order. The people's occupations and residences could only be passed on from their parents and remained unchanged. People could not have travel without the permission of *Ryoshu*. Any farmland transaction other than inheritance, such as buying

[2] Vegetation transition is used for both agriculture and forestry. Thus, forestry is sometimes categorized as part of the agricultural industry in Japan (Source: Sekioka, H., *Chishiki Zero Karano Ringyo Nyumon* (Basic Knowledge on Forestry), (Japan: Ieno Hikari Kyokai), 2016, p. 19).

[3] Some agricultural products are inedible. A typical example is the thread from a silk cocoon, which is also in competition (in commercial use) with new types of materials such as chemical fibre and glass fibre.

and selling, lending and borrowing, and pawning, were prohibited in principle. Farmers were not allowed to grow crops other than rice unless they received special permissions from *Ryoshu*.

Ryoshu Sei was efficient in enforcing collaboration among farmers. Indeed, agricultural production increased nearly 1.5 times during the first 100 years of the Tokugawa Shogunate period.[4] However, *Ryoshu Sei* had a serious flaw: farmers had little incentive to try new types of technologies and varieties. In other words, as long as farming technologies remained the same, *Ryoshu Sei* worked well. However, in the late nineteenth century, although new technologies emerged, such technologies were not implemented because of the inflexibility of *Ryoshu Sei*.[5] Forestry as a natural resource was a common capital good among villagers. Since *Ryoshu Sei* ensured farmers in a village had joint use of forestry beyond the current generation, farmers used forestry resources sustainably with a long-term view. Today, many environmental historians recognize that the management and usage of forestry resources in the Tokugawa Shogunate period was an ideal case of best practice in terms of environmental protection.[6]

The Meiji Restoration (1868) and Removal of Feudal Regulations

Commanders of warships that arrived in Japan during the period of strict national seclusion strongly demanded that the Tokugawa Shogunate open up Japan to the West. Although the Tokugawa Shogunate first attempted to repulse these demands, it failed eventually. In 1858, it concluded the Shimoda Treaty with the US government, which stipulated that Japan should open up to the US. This was just the beginning, and the Tokugawa Shogunate went on to conclude similar treaties with the Dutch, Russian, and British governments. This policy change jolted the Japanese society and disrupted the regime of the Tokugawa Shogunate. As a result, a new regime, the Meiji government, took power in 1868 (Meiji is the Japanese

[4] Teruoka, S., *Nihon no Nogyo Hyakugoju Nen* (Japanese Agriculture 150 Years), (Japan: Yuhikaku), 2003, p. 7.

[5] Hayami, Y., and S. Yamada, *The Agricultural Development of Japan*, (Japan: University of Tokyo Press), 1991, pp. 61–62.

[6] For example, see Saito, O., *Kankyo no Keizaishi* (Economic History of Environment), (Japan: Iwanami Shoten), 2014, pp. 38–42, and Richards, J. F., *The Understanding Frontier: An Environmental History of the Early Modern World*, (California: University of California Press, Berkeley), 2003, p. 622.

name for the imperial rule lasting 45 years from 1868 to 1912). The 1868 event is called the Meiji Restoration, which is seen as the epoch of Japan's full-scale assimilation with the West. The Meiji Restoration provided the platform for Japan's modern economic growth. The Meiji government adopted economic modernization measures in quick succession in the following three ways.

The first was the introduction of modern technologies from Western countries. For example, the government ran model factories where advanced Western technologies were implemented and showcased. Some technologies did not work well in Japan because of shortage of technicians and underdevelopment of social infrastructure. However, after trials and errors, these technologies were transformed based on the local conditions and features of the Japanese factories, which eventually enhanced their productivity. The second way was making investments in social infrastructure such as roads, railways, and port facilities. This was in sharp contrast to the policy of Tokugawa Shogunate, which prohibited *Ryoshu* from constructing large-sized bridges because the Shogunate wanted to control the traffic flow of people and materials. In addition, the Meiji government developed intangible social infrastructure such as foreign exchange banks and the telecommunication system.

The third, and the most important, aspect was the liberalization of market-driven economic activities through the removal of feudal regulations. One most important reform was the Land Tax Revision, which took place from 1873 to 1881. The revision changed the feudal tax, which was levied in proportion to the quantities of crops harvested, and it brought about the institution of the modern land tax system, which was based on the value of land. For this reform, the government issued a land certificate to the landowner of each parcel of land. Each land certificate identified the owner of the farmland and the evaluated value of the land. The owner was obliged to pay a yearly tax amounting to as much as 3% of the evaluated value. The prohibition on farmland leasing, pawning, and trading was entirely removed. Farmers were free to produce any crops according to their choice. In addition, the Meiji government granted the people freedom of occupation, farmland transaction, agricultural production, and transportation. After the Meiji Restoration, specialization and division of labour developed in the Japanese economy. Accordingly, the economic activities of coal and wood collection were separated from agriculture and became the coal mining and forestry industries respectively.

Structural Changes in the Primary Sector After the Meiji Restoration

The Land Tax Revision ensured that the Meiji government would have stable tax revenues. In addition, this revision introduced free competition into the farmland market. In particular, wealthy families accumulated farmland by purchasing land certificates. However, this does not mean that wealthy families operated more farmland by themselves. At that time, because of underdevelopment of agricultural machinery and pesticides, the optimal farming size of agricultural plots was limited to around 1.0 hectare. Thus, wealthy families also farmed land of around 1.0 hectare by themselves. The remaining farmland was rented out to tenant farmers. Such wealthy families are called "self-farming landowners," and they earned agricultural income from self-farming activities and collected rent from tenant farmers.

Such self-farming landowners not only had rich information on local agriculture but were also held in high social esteem in their hometowns because of their political and economic leadership within the community. As such, they were actively involved in the formation of common agricultural capital for projects such as joint irrigation facilities. They also functioned as providers of a socio-economic safety net for poor families. For example, they showed consideration for villagers by reducing rent during the bad harvest years. Thus, the rural society experienced favourable socio-economic conditions, as villagers collaborated with each other under the leadership of self-farming landowners. However, this situation gradually became impossible to maintain as the Japanese economy shifted from the light industrialization period to the heavy industrialization phase, discussed in the following two sections.

Coal mining workers also formed a unique community system, called *Hanba Sei*.[7] While the original meaning of *Hanba* is a simple building that provides meals and lodging to workers, the word *Hanba* was also used for a unit of coal miners. Each *Hanba* had a leader, called *Oyakata*. Usually, a coal mining company did not enter into direct employment contracts with individual coal miners. Instead, it paid money to *Oyakata*, who took the responsibility of labour management, including recruitment and allocation

[7] Details of *Hanba Sei* are discussed in Yano, M., T. Tanji and M. Kuwabara, *Sekitan no Kataru Nihon no Kindai* (Coal and Modernization in Japan), (Japan: Soshi-ete), 1978, pp. 145–152.

of jobs among workers. *Oyakata* was also in charge of providing a financial safety net to miners. For example, *Oyakata* provided money to miners when they fell ill. As labour management tools and social safety nets were underdeveloped in this period, *Hanba Sei* is considered to have been a useful system.

For forestry, the Meiji government's removal of feudal regulations brought disorder in resource use. Responding to the surge in the demand for wood, the total amount of timber felled increased immediately without consideration for the sustainability of long-term benefits. As a result, destruction of the forests reached a critical level.[8] The Meiji government was aware of the necessity of protecting forests from overcutting. Therefore, the government implemented the National Forestry Regulations in 1871 and the Temporary Ordinance for National Forestry Survey in 1876. However, these legal frameworks were not workable because they did not provide sufficient incentives for forestry users to follow the regulations.

The Dawn of Heavy Industrialization

After the Meiji Restoration, productivity in the primary sector improved. Demand for domestic wood and food also increased. However, the demand for domestic coal did not increase as much because wood was more popular among domestic consumers. As a result, for nearly 15 years after the Meiji Restoration, about half the coal produced was exported overseas.[9] After the Russo-Japanese War (1894–1895), Japan moved from light to heavy industrialization period. There was development of large-sized factories such as shipyards and ironworks that needed high amount of energy. Accordingly, the demand for domestic coal increased sharply, and coal exports diminished.[10]

In this period, railroading and forestry developed in tandem with the coal mining industry. Tramcars were introduced for carrying coal miners and equipment to mining sites and carrying out low-grade coal from

[8] Yamaguchi, A., *Shinrin Shigen no Kankyo Keizaishi* (Environmental History of the Forestry Resource), (Japan: Keio University Press), 2015, p. 5.

[9] Yada, T., *Sekitan Sangyo Ron* (Studies on the Coal Mining Industry), (Japan: Harashobo), 2014, p. 16.

[10] Japan became a coal importer in 1923 (Source: Sugiyama, S. and T. Ushijima, 2012, *Nihon Sekitangyo no Suitai* (Decline of the Coal Mining Industry), (Japan: Keio University Press), 2012, p. 8).

mining sites to coal dressing sites. In addition, the use of railway was popular for carrying coal from coal dressing sites to coal consumption areas. Wood is an important material for use in both coal mining industry and railroading. Wood is used for pit props and construction of railway sleepers. The logistics of moving goods still depended on energy fuels. In the case of trains, coal is consumed by locomotive engines as a form of fuel. In addition to the heavy industry, the number of paper factories, which created a strong demand for coal for fuel and wood for pulp, increased in this period.[11] Paper products were shipped to consumers by railways. This was another example of synergy between the coal mining, forestry, and railroading industries.

Impact of Heavy Industrialization on Human Relationships in the Primary Sector

To operate heavy industry factories, a large supply of trained workers is imperative. The discipline required to work in close coordination with many fellow workers in a manner consistent with the rhythm of machine operation, as dictated by the inorganic time of a clock and under the command of supervisors, is not easy to acquire for a worker. This is especially for a peasant who is used to farming on his own land, in accordance with the traditional rhythms of nature and implicit collaboration with family members and close neighbours.[12] Such traditional rhythm is called "peasant time."[13] For workers used to working according to peasant time, developing the aptitude to work in the factory system involves a change in their broad social value system or culture. If workers who cannot change their culture, that is, if workers who continue to stick to peasant time are employed in a factory, their frequent absenteeism would damage the working of the factory system, resulting in a serious loss in production efficiency.

School education at the primary and secondary levels represents a highly effective apparatus for embedding aptitude for the factory system among the youth in traditional agrarian societies. By being forced to come

[11] Fuji Paper Kushiro Factory, which was one of the biggest paper factories in the pre-war period, started its operations in 1920.

[12] Please refer to the following for a full treatment: Thompson, E. P., "Time, Work-Discipline, and Industrial Capitalism," in *Past and Present*, 38 (December), 1967: 56–97.

[13] Please refer to the following for a full treatment: Smith, T. C., *Native Sources of Japanese Industrialization, 1750–1920*, (Berkeley and Los Angeles: University of California Press), 1988.

to school at a prescribed time and study for the duration of class hours together with classmates under the instruction of teachers, children are regimented to follow the discipline of collective action. School-wide activities, such as morning assemblies, entrance and graduation ceremonies, and sports days, provided them opportunities to become accustomed to larger collective actions under hierarchical commands. Basic literacy and arithmetic, as well as the elementary knowledge of sciences, acquired by students at primary- and secondary-level schools add much to the efficiency of their eventual labour work by ensuring they gain a better understanding of production manuals and instructions given to them when they are employed at factories.

The demand for schools that can mass-produce labourers with the aptitude for the factory system is especially large in economies that have newly introduced the factory system as an inseparable process of technology borrowing. This is evidenced by the faster and more complete diffusion of elementary education systems in late-industrializing countries such as Germany and the US as compared with the UK. Similarly, the Meiji government launched an ambitious project through the School System Rule (*Gakusei*) of 1872 to propagate the compulsory elementary education system throughout the nation. This proclamation, passed only four years after the Meiji Restoration, had a famous preface announcing its goal that "there shall not be a single household in any village, which does not let children study in school." This target was virtually achieved in the following four decades, as school enrolment ratios of elementary school–aged children rose from 28% in 1873 to 81% in 1900, and climbed to 98% in 1910.[14]

Hanba Sei can be seen as a style of creating labour contracts in the coal mining industry within a society wherein peasant time was dominant. However, as discussed above, because of heavy industrialization, the aptitude of Japanese workers changed even in the primary sector. As a result, *Hanba Sei* became unpopular in this period. Instead, a majority of coal miners entered into direct employment contracts with coal mining companies.

Human relationships in farmers' communities also changed because of heavy industrialization. As the economic concerns of wealthy families changed, they began investing in new factories in urban areas to generate

[14] Japanese Ministry of Education, *Wagakuni Kyoiku no Ayumi* (History of Japanese Educational Development), (Tokyo: Meiji Tosho Shuppan), 1967.

greater financial returns and became deeply engaged in the management of heavy industrial businesses. Even though local rich families moved to urban areas, they usually kept their farmland property as part of their portfolio. These new types of landowners are called "absentee landowners." In contrast to self-farming landowners, absentee landowners neither had good information on local agriculture nor took leadership positions in forming social overhead capital in rural areas. As the heavy industrial sector in urban areas grew, the proportion of self-farming landowners decreased and that of absentee landowners increased.

There existed conflicting interests between absentee landowners and tenant farmers. For example, absentee landowners often demanded increased rents from tenant farmers, leading to strong resistance and often fierce village-wide disputes, termed "peasant disputes." Peasant disputes started in the second half of the 1910s in western Japan, where heavy industrialization first began and then spread nationwide during the 1920s. At their peak, the number of peasant disputes reached about 7000; consequently, farmers could not concentrate on farming.[15]

In order to rebuild community order among farmers, the government introduced a new organization, called *Sangyo Kumiai*. The literal meaning of *Sangyo Kumiai* is cooperative. Cooperatives in European and North American countries are founded and operated based on citizens' autonomy and kept independent from the government. When *Sangyo Kumiai* started in 1900, it was modelled after the cooperatives in European and North American countries. However, in 1932, the government intervened in *Sangyo Kumiai* and reformed it into a de facto sub-administrative body. With strong ties to the municipal government, *Sangyo Kumiai* was responsible for drawing up and carrying out self-salvation plans in the farming community.

The Japanese government took a similar approach for the forestry industry. In 1907, the government launched a new framework, called *Shin-rin Kumiai*, for those who are engaged in forestry. The literal meaning of *Shin-rin Kumiai* is "forestry cooperative." *Shin-rin Kumiai* was (and still is) in charge of setting up and carrying out the forestry use project. Participation in the project was made compulsory upon consent of more than two-thirds of the landowners owning more than two-thirds of the forest in the districts concerned. In Japan, a forest area was (and still is)

[15] Teruoka, S., *Nihon no Nogyo Hyakugoju Nen* (Japanese Agriculture 150 Years), (Japan: Yuhikaku), 2003.

usually shared by many small-sized landowners. This made (and still makes) coordination of interests among landowners difficult. *Shin-rin Kumiai* is the government's strong-handed policy measure to enforce collaboration under the guidance of the local authority. That is, the local government invites the major landowners of a forestry area and encourages them to establish *Shin-rin Kumiai*. As such, *Shin-rin Kumiai* works as a coercive measure to organize forestry landowners for a forestry project.[16]

Hardships Faced in the Primary Sector During Pre-war Japan's Heavy Industrialization

In the 1900s and 1910s, Japan's heavy industry achieved good performance. This is mainly because European countries, which were major exporters of heavy industry commodities at that time, reduced production during the turmoil of World War I. However, after the economic recovery of European countries, the Japanese heavy industry faced hardships. Japan experienced the Reactive Depression in 1920, one of the biggest depressions in Japanese history, followed by chronic economic stagnation for more than a decade. Unlike light industry, heavy industry is associated with a strong-scale economy. Thus, once a large-scale factory is built, it is difficult to downsize the operation even if demand becomes weak. In addition, the government's commitment to the business society becomes much stronger because heavy industry requires large public investment in industrial infrastructure, such as high-level vocational training institutes and industrial complex sites.

Indeed, the Japanese government provided various forms of support to the heavy industry. As part of the support, the government attempted to suppress the prices of food commodities and raw materials (food prices are important for keeping the wage rate of factory workers at a low level). For example, the government removed the wood tariff in 1920. As a result, low-priced wood from the US flooded into Japan.[17] In addition, wood

[16] Nagata, S., *Rinseigaku Kogi* (Lectures on Forestry Policy), (Japan: University of Tokyo Press), 2015, p. 47.

[17] In 1924, the government reduced the tariff on imported wood as an emergency measure to address the wood shortage after the Great Kanto Earthquake. As a result, wood imports from the US (including British Colombia) increased from 0.77 million goku (1 goku is equivalent to 0.278 cubic metres) in 1920 to 9.93 million goku in 1924 (Source: Yamaguchi, A.,

production in Sakhalin increased and the wood flooded into Japan as part of the Japanese government's colonial policy.[18] In the latter half of the 1920s, the share of imported wood was about 24–25% every year.[19]

The Japanese government introduced an aggressive programme for the import of food from its colonies, called *Sanmai Zoshoku Keikaku* (Rice Production Development Plan). Under *Sanmai Zoshoku Keikaku*, the Japanese colonial governments in Taiwan and Korea invested in irrigation and water control and in research and extension, in order to develop and defuse high-yielding Japanese rice varieties adapted to the local ecology. The success of this effort created a tremendous rice surplus that flooded the Japanese market. In the 20 years from 1915 to 1935, net imports from Korea to Japan rose from 170,000 to 1,212,000 metric tons per year, and net imports from Taiwan rose from 113,000 to 705,000 metric tons.[20] As a result, net import of rice rose from 5% to 20% of domestic consumption.[21] Accordingly, rice prices dropped sharply. In addition, the international price of silk also started declining because of the development of synthetic fibre. Thus, a majority of farmers fell into a miserable situation called the "Agricultural Depression," which stirred up serious social unrests nationwide.

Rural poverty resulted in massive outflow of cheap labour from rural areas to the coalfields.[22] Unlike for food and wood, the government was not active in boosting coal imports. This is because Japan was endowed

Shinrin Shigen no Kankyo Keizaishi (Environmental History of the Forestry Resource), (Japan: Keio University Press), 2015, p. 45).

[18] Forestry in Sakhalin suffered substantial damage from insects in 1919. In response, the Japanese colonial government in Sakhalin cut down the damaged wood as a preventative measure to protect the insect-free forestry areas. This damaged wood was shipped within Japan and sold at cheap prices domestically. In addition, although overcutting of forestry reached a critical level in Sakhalin, the Japanese colonial government did not employ effective measures to prevent it (Source: Yamaguchi, A., *Shinrin Shigen no Kankyo Keizaishi* (Environmental History of the Forestry Resource), (Japan: Keio University Press), 2015, p. 46–48).

[19] Yamaguchi, A., *Shinrin Shigen no Kankyo Keizaishi* (Environmental History of the Forestry Resource), (Japan: Keio University Press), 2015, p. 47.

[20] Hayami, Y., and S. Yamada, *The Agricultural Development of Japan*, (Japan: University of Tokyo Press), 1991, p. 79.

[21] Hayami, Y., and S. Yamada, *The Agricultural Development of Japan*, (Japan: University of Tokyo Press), 1991, p. 79.

[22] Yano, M., T. Tanji and M. Kuwabara, *Sekitan no Kataru Nihon no Kindai* (Coal and Modernization in Japan), (Japan: Soshi-ete), 1978, pp. 147–149.

with relatively rich coal resources.[23] Since the coal mining industry is labour intensive, labour costs shared a critical part of the production cost of coal.[24] In sum, the biggest concern for the government was promoting the secondary sector. To do so, the government expected the primary sector to provide food, fuel, and materials at low prices.

Protection for the Primary Sector During Post-war Japan's Heavy Industrialization

Japan entered a war economy in the 1930s and trade barriers such as import tariffs increased in the early 1930s.[25] Isolated from the international society, Japan suffered from serious shortage of food, energy, and wood. To cope with such hardship, the Japanese government ceased the market mechanism and reformed itself into a controlled economy. The Pacific War ended in 1945 but the Allied Forces occupied Japan until 1952. The General Headquarters of the Allied Forces (GHQ) exercised strong leadership in the introduction of democracy to Japan. As part of the GHQ's democratization policies, *Nogyo Kai*, which was scaled up from *Sangyo Kumiai* in 1943, was dissolved in 1947. *Shin-rin Kumiai* was also converted into a democratic system, which forestry landowners voluntarily established and operated. However, the GHQ continued to control the economy because the shortage of necessities such as food, fuel, and wood was still serious. Increasing the primary sector's products was the top priority.

The GHQ announced a broad policy of abolishing the controlled economy in the end of 1949. Based on this policy, market economy gradually resumed. Heavy industrialization restarted, too. Then, in contrast to the

[23] Yano, M., T. Tanji and M. Kuwabara, *Sekitan no Kataru Nihon no Kindai* (Coal and Modernization in Japan), (Japan: Soshi-ete), 1978, p. 20.

[24] Since coal mining is a labour-intensive industry, keeping a sufficient number of miners is one of the biggest problems when lowering the coal production cost. When the labour shortage became serious in the Pacific War period, the government forced people in Japanese colonies and prisoners of war to work as involuntary labourers at coalfields. (Source: Yada, T., *Sekitan Sangyo Ron* (Studies on the Coal Mining Industry), (Japan: Harashobo). 2014, pp. 67–68).

[25] The nominal protection rate for imported rice increased from 11% in 1923–1927 to 45% in 1933–1937 (Source: Hayami, Y., and Y. Godo, *Nogyo Keizai ron* (Agricultural Economics), (Japan: Iwanami Shoten), 2002, p. 147).

period of the controlled system, Japan's primary sector became a fetter for economic growth. Under the international free trade system, it became possible for Japan to import coal, food, and wood at lower prices from all over the world. In addition, usage of oil for fuel and iron and plastics for construction materials increased rapidly after the Pacific War. As a result, demand for domestic coal and wood became weak. Indeed, domestic coal and wood production hit their respective peaks in 1961 and 1957 and then showed a sharp downward trend. Demand for domestic agricultural products also stagnated because of the westernization of eating habits and inflow of cheap crops from the US. The total amount of domestic grain production kept showing a downward trend after 1960.

The so-called processing trade, whereby Japan imports raw materials and exports manufactured goods, was the major strategy of post-war Japan's heavy industrialization. Production resources such as labour and capital moved from the primary sector, which produces raw materials, to the secondary sector, which produces manufactured commodities. Under such a situation, the government created a new development framework and outline for each industry of the primary sector under the following: the Agricultural Basic Law of 1961, the Outline of Coal Policy of 1962, and the Forestry Basic Law of 1964. These three laws shared the same basic view as the following two points:

(1) The government should change its traditional stance of policymaking, which focuses on the promotion of domestic production, to a new priority, which focuses on the improvement of the economic conditions of those involved in the industry.
(2) The government admitted that the overall demand for the primary sector's domestic products will not increase anymore and presented a new industry strategy:[26] reduction of the production cost by implementing advanced production technology and concentrating on certain types of commodities.

In the Outline of Coal Policy of 1962, under the slogan of "scrap and build," the government suggested that coal mining companies distinguish their mining pits into two groups: scrap group and build group. The build group refers to pits that have large sizes, are easy to mine, and produce

[26] Annual statistics on coal production are available at the Sekitan Kogyo Go-rika Seisakushi Kenkyu Kai (1990). The government's estimates regarding the food balance sheet are available at http://www.maff.go.jp/j/zyukyu/fbs/. The government's estimates regarding the wood balance sheet are available at http://www.maff.go.jp/j/tokei/kouhyou/mokuzai_zyukyu/

high-quality coal. The remaining pits would be in the scrap group. The government classified scrap group pits for closure one after the other. However, there were two major problems in closing the pits. One is the bad debt of coal mining companies, which had accumulated since the so-called Coal Depression of 1958. Thus, the government provided various schemes of financing public funds to relinquish these bad debts. The other problem was unemployment of miners. Indeed, since the pre-war period, almost every time a coal mining company presented a plan of closing a coal mining pit, miners, who would lose their jobs, showed strong resistance to the plan. Thus, the government prepared various relief measures to address unemployment in cities whose coal mining pits were being closed. These measures mitigated (although they did not entirely cease) coal miners' movements against closing mine pits.

The government was of the view that pits of the "build group" would conquer imported coal and oil if coal companies installed advanced machinery in mining cites. Based on this view, the government provided subsidies for investing in advanced machinery. However, the build group mining pits did not improve their profitability because simple mechanization was not sufficient to reduce production costs. Instead, a complete restructuring of the management system was necessary.[27] As a result, even the build group pits entered the list of closing pits. While 909 coal mining pits were open in 1955, only 73 pits remained open by 1970.[28] In sum, despite consuming an enormous amount of public funds, the coal mining industry diminished to a negligible level in the Japanese economy by the end of the 1960s.

The structure of the agricultural society changed entirely following drastic agricultural land reforms from 1946 to 1949, whereby farmland owned by absentee landowners was transferred by force to tenant farmers. This was recognized as one of the most significant democratization policies of the Allied Forces and resulted in a large number of former tenant farmers becoming owner farmers. They became organized into a national network of agricultural cooperatives known as *Nokyo,* which was established in 1947. *Nokyo* provided not only agricultural services such as joint shipments but also a wide range of non-agricultural services such as

[27] Sugiyama, S. and T. Ushijima, *Nihon Sekitangyo no Suitai* (Decline of the Coal Mining Industry), (Japan: Keio University Press), 2012, pp. 218–219.

[28] Sugiyama, S. and T. Ushijima, *Nihon Sekitangyo no Suitai* (Decline of the Coal Mining Industry), (Japan: Keio University Press), 2012, pp. 14–16.

supermarkets, gasoline stations, banks, funeral services, and hospitals. *Nokyo* was a de facto sub-administrative body, too. Moreover, agricultural subsidies were distributed to farmers through *Nokyo*, which also ensured coordination among farmers and promoted the formation of common agricultural capital. *Nokyo* functioned as a strong vote-receiving mechanism as well. *Nokyo* was one of the most influential supporters of the Liberal Democratic Party (LDP), which was in power continuously from 1955 to 1993 (when it lost to the opposition for a brief few months) and therefore until today (except for 2010–2013, when the political opposition won again). Farmers' votes for LDP politicians were essential for the party to stay in power during that time. Thus, LDP implemented strong income redistribution policies for farmers.

In 1961, the government enacted the Agricultural Basic Law, a national charter for agriculture. The law declared that it was the government's responsibility to raise agricultural productivity and thereby close the gap between the income and the welfare of farm and non-farm people. One measure identified as necessary for this purpose was the so-called selective expansion, which provided encouragement to expand selectively the production of agricultural commodities in response to the changing demand structure. The government designated fruits, meat, and milk as "promising commodities" whose consumption would keep growing. The government also promoted large-sized and mechanized farming to improve agricultural productivity. For "selective expansion" and mechanization, the government started various subsidy programmes for farmers.

Upon receiving subsidies, the income level of farmers increased faster than that of non-farmers. However, the government's projection of food consumption turned out to be incorrect. Even for promising commodities, consumers' demand did not grow as the government had expected. In addition, the government's encouragement of mechanization often resulted in overinvestment, which resulted in bad debts for farmers. To reduce such debts, the government implemented rescue loans financed by the national treasury. In sum, agricultural subsidies were consumed to create an excess supply of agricultural commodities, but there was no improvement in agricultural productivity.

The government's scenario in the Forestry Basic Law was summarized as follows:

> Provision of large-sized public loans to managers of the forestry industry → increase in productivity → enlargement of the unit of forestry management → modernization of forestry management to increase productivity

This scenario focuses on only managers and workers in the forestry industry. Here, a big flaw was that the behaviour of forestry owners, which was (and still is) critical of how to use Japanese forestry, was ignored.

In Japan, the majority of forestry owners live in urban areas and are not concerned with the management of forestry. After the Pacific War, the government became active in constructing long-distance public roads in mountainous areas by procuring forest lands from owners at prices higher than their market value. In addition, the asset tax was (and still is) low for forestry owners. As a result, the majority of forestry owners were (and still are) more interested in inviting public construction projects than in improving forestry management. Ironically, *Shin-rin Kumiai* benefits those who have ownership of forestry but are unfamiliar with forestry management. The requirement of membership in *Shin-rin Kumiai* is to have ownership of forestry. By hiring forestry workers, *Shin-rin Kumiai* takes care of the members' forests. (A forestry worker cannot be a member of *Shin-rin Kumiai* unless he or she holds ownership of forestry.)

In the 1950s and 1960s, the government promoted conversion of natural forests to manmade forests by providing subsidies. A major recipient of these subsidies for the management of manmade forests was (and still is) *Shin-rin Kumiai*. Thus, relying on *Shin-rin Kumiai,* many landowners became less interested in the details of forestry management and simply waited (and still wait) for an opportunity to sell their land to the government for public construction projects. Today's experts conclude that the forestry policy implemented after the Forestry Basic Law simply made forestry owners more dependent on subsidies and did not improve the management of the forestry industry.[29]

Another flaw in the government's forestry policy was its misplaced belief that Japanese consumers appreciate Japan's knotless woods so highly that domestic forestry can remain profitable even after liberalization of the wood trade. This misconception was called "the myth of domestic knotless wood."[30] Based on this mistaken belief, the government encouraged

[29] Mochida H. and K. Endo, *Ringyo Kozo Mondai Kenkyu* (Studies on Industrial. Structure of Japanese Forestry), (Japan: Japan Forestry Investigation Committee), 2015, p. 20.

[30] Murao fiercely criticizes such misunderstanding as the "propagation of a cult" (Source: Murao, K., *Machigai Darakeno Nihon Ringyo* (Mistakes of the Japanese Forestry), (Japan: Japan Forestry Investigation Committee), 2013, p. 53).

production of knotless wood.[31] In reality, however, insufficient timber drying was the real reason for the shrinkage of the demand for domestic wood. Throughout the 1960s, imported wood and new types of construction materials such as plastics replaced the demand for domestic wood.

Conclusion and Discussion

As discussed above, the historical performances of the three industries in the primary sector have several similarities. The primary sector showed good performance from the proto-industrialization period to the light industrialization period. However, when Japan faced hardships during the period of pre-war heavy industrialization, the government supported the secondary sector by suppressing the prices of primary sector commodities. For example, the government promoted the import of low-priced food and wood, which resulted in rural poverty. The affected farmers moved to mining sites to work as cheap labourers. While rural poverty stirred up social unrest, the government did not redistribute income sufficiently to rural areas because democratization in pre-war Japan was still limited and the political power of the primary sector was weak.

As heavy industrialization restarted in the post-war period, the political situation of the primary sector differed from that in the pre-war period because of democratization during the Allied Forces' occupation. The three industries in the primary sector kept losing market share in the post-war heavy industrialization period, and thus all three received large subsidies. According to the government's official view, these subsidies were for improving the productivity of the primary sector. In reality, however, these subsidies should be regarded as income transfer to the primary sector, and they did not contribute to the improvement of productivity.

However, this does not mean that the subsidies were useless. Considering that the cross-sector income gap in the pre-war period produced social unrest, which fostered (left-wing) militarism, income redistribution to the primary sector could have contributed to economic growth by preventing social instability. Nevertheless, it should be noted that when the government implemented the income redistribution policy for the primary sector,

[31] Please refer to: Murao, K., *Machigai Darakeno Nihon Ringyo* (Mistakes of the Japanese Forestry), (Japan: Japan Forestry Investigation Committee), 2013, pp. 163–167, and Ogi, T., Kokusanzai wa Naze Urenakattanoka (The Reasons for Stagnation in Japanese Forestry), (Japan: Japan Forestry Investigation Committee). 2009, p. 116.

it presented rosy scenarios such as "scrap and built," "selective expansion," and "the myth of domestic knotless wood." Later, these scenarios turned out to be wrong. It was not rare to see those who received public loans based on these hypothetical scenarios suffer from bad debts. While terms such as "scrap and built" and "selective expansion" may have been necessary due to the politics involved in preparing subsidies, they brought unnecessary confusion to the primary sector.

The discussion in this chapter shows that, in order to gain a comprehensive understanding of the economic history of the coal mining industry, comparative studies with other industries in the primary sector are useful. However, as discussed in the beginning of this chapter, such comparative studies have not been conducted sufficiently in the Japanese academic society. The author hopes that this chapter is the first of many economic studies on the coal mining industry and the overall primary sector. The chapter provides a useful context and background of the onset of modernization and industrialization in Japan. And within that backdrop, there was a gradual transition from the agricultural to the industrial sector. Industrialization then progressed from light to heavy industries (of which coal was one of the big four, together with shipbuilding, steelmaking, and equipment/car manufacturing). Along with industrialization, workers were disciplined, rationalized, and mass-educated to increase productivity to maximize output. This chapter ends with the onset of postwar industrialization, after having surveyed how the coal mining industry was eventually outcompeted and diminished by imports. The next chapter continues the narrative chronologically and picks up from this chapter to analyse the factors that led to Japan's domestic coal mine decline and policies to manage the transition, especially for laid-off workers.

Bibliography

Hayami, Y., and Y. Godo, *Nogyo Keizai ron* (Agricultural Economics), (Japan: Iwanami Shoten), 2002.

Hayami, Y., and S. Yamada, *The Agricultural Development of Japan*, (Japan: University of Tokyo Press), 1991.

Japanese Ministry of Education, *Wagakuni Kyoiku no Ayumi* (History of Japanese Educational Development), (Tokyo: Meiji Tosho Shuppan), 1967.

Mochida H. and K. Endo, *Ringyo Kozo Mondai Kenkyu* (Studies on Industrial. Structure of Japanese Forestry), (Japan: Japan Forestry Investigation Committee), 2015.

Murao, K., *Machigai Darakeno Nihon Ringyo* (Mistakes of the Japanese Forestry), (Japan: Japan Forestry Investigation Committee), 2013.
Nagata, S., *Rinseigaku Kogi* (Lectures on Forestry Policy), (Japan: University of Tokyo Press), 2015.
Ogi, T., *Kokusanzai wa Naze Urenakattanoka* (The Reasons for Stagnation in Japanese Forestry), (Japan: Japan Forestry Investigation Committee). 2009.
Richards, J. F., *The Understanding Frontier: An Environmental History of the Early Modern World*, (California: University of California Press, Berkeley), 2003.
Saito, O., *Kankyo no Keizaishi* (Economic History of Environment), (Japan: Iwanami Shoten), 2014.
Sekioka, H., *Chishiki Zero Karano Ringyo Nyumon* (Basic Knowledge on Forestry), (Japan: Ieno Hikari Kyokai), 2016.
Sekitan Kogyo Go-rika Seisakushi Kenkyu Kai, *Sekitan Kogyo Go-rika Seisakushi* (History of the Streamlining Policy for the Coal Mining Industry) (Japan: Sekitan Kogyo Go-rika Seisakushi Kenkyu Kai), 1990.
Shiga, K., *Shinrin Kanri Ron* (Forestry Management), (Japan: Japan Forestry Investigation Committee), 2016.
Shimanishi, T., *Nihon Sekitangyo no Sengoshi* (History of the Coal Mining Industry), (Japan: Keio University Press), 2011.
Smith, T. C., *Native Sources of Japanese Industrialization, 1750–1920*, (California: University of California Press, Berkeley and Los Angeles), 1988.
Sugiyama, S. and T. Ushijima, *Nihon Sekitangyo no Suitai* (Decline of the Coal Mining Industry), (Japan: Keio University Press), 2012.
Teruoka, S., *Nihon no Nogyo Hyakugoju Nen* (Japanese Agriculture 150 Years), (Japan: Yuhikaku), 2003.
Thompson, E. P., “Time, Work-Discipline, and Industrial Capitalism,” *Past and Present*, 38 (December), 1967: 56–97.
Yada, T., *Sekitan Sangyo Ron* (Studies on the Coal Mining Industry), (Japan: Harashobo). 2014.
Yamaguchi, A., *Shinrin Shigen no Kankyo Keizaishi* (Environmental History of the Forestry Resource), (Japan: Keio University Press), 2015.
Yano, M., T. Tanji and M. Kuwabara, *Sekitan no Kataru Nihon no Kindai* (Coal and Modernization in Japan), (Japan: Soshi-ete), 1978.

CHAPTER 3

The Miike Coal Mine and Omuta City: A Case Study of the Largest and Last Mine's Closure in Japan

Introduction

This chapter continues the narrative of coal mine developmental history from the previous chapter (Chap. 2) by Godo. While Godo detailed the rise and emergence of the heavy industry of coal mining and discussed mindset changes, infrastructure developments, skills training/education, and the transition of an agricultural economy to an industrializing one, this chapter details how the process reached its peak and then faced closure in another phase of Japanese economic development in the post-war period. To demonstrate the macro changes taking place in the overall general economy, the Japanese sociologist Naoko Shimazaki, an experienced scholar in research on coal mine closure, utilized micro case studies of Miike and Omuta to indicate the layered complexities of coal mine closures, especially at the individual mine level. As a sociologist, Shimazaki's chapter is also a people-centred narrative/story unveiling a tale of individuals, families, and entire towns affected by the closures and the very real emotive realities that they face when the local mining economy dissipated. It continues the historical narrative from Godo's macro developmental economic analysis but with local empirical case studies based on sociological perspectives.

Naoko Shimazaki's chapter is centred on the Miike Coal Mine. There are at least three significant points in studying this coal mine. First, it is the largest coal mine in Japan that has the highest quality of coal in Japan. The

T. W. Lim et al., *Coal Mining Communities and Gentrification in Japan*, https://doi.org/10.1007/978-981-13-7220-9_3

government also tried to make the coal mine extraction sustainable. Because of its scale, the Miike Coal Mine was economically important and therefore politically influential. Because of its political significance, labour relations and coal strikes in Miike also garnered the most attention, particularly from the mass media. In the 1960s, the largest industrial incident involving a coal mine to date also occurred in Miike. All these features made Miike first amongst equals in terms of academic and research significance in the study of Japanese coal mines.

Second, it is simultaneously the last coal mine to close in Japan. While, technically speaking, Kushiro and Ikeshima coal mines were the last to close in Japan, Miike's closure was far more significant as its scale was the largest amongst the three. The third significance is the chapter's contribution to existing literature, as there is no English-language literature at the moment that studies Miike's closure in detail and within a conceptual/theoretical framework. Facing the coal industry's decline in 1980s, Miike and Omuta City sought to transform the perception and imagery of the region from a "coal mining town" to a "town with coal mining" (i.e. an industrial area that included coal mining), and so avoiding the negative label of a former "town where there used to be coal mining." This chapter addresses the question of whether this transformation was successful.

Overview of the Coal Industry in Japan: A Short Survey of the Japanese Coal Industry

This chapter begins with an overview of individual coal mine histories. In Japan, the first discovery of coal was "combustible stone" at Miike Inariyama in 1469 and at Chikuzen Onga in 1478, both locations found in Kyushu. Later, during the Edo Period (1603–1868), coal was first extracted by the Kyushu clans of Miike, Nabeshima, and Tagawa. At that time, coal was used mainly as the source of energy for salt production in the Setouchi area. At the end of the Edo Period (1603–1868), the Edo government brought about the end of Japan's period of isolation (known as the policy of *sakoku* in Japanese language). Hakodate and Shimoda ports were opened up in order to supply coal for American ships which aimed to trade with China and to engage in whale fishing. Also, the Takashima Coal Mine in Kyushu and the Kayanuma Coal Mine in Hokkaido were opened. The Takashima Coal Mine was the first modernized coal mine in Japan to adopt British techniques of extracting coal.

During the Meiji Period (1868–1912), mining production and a fundamental system of management were established in Japanese coal mines. At the time of the 1873 Mining Industry Act, the main coal-producing mines were the Takashima Coal Mine and the Miike Coal Mine. As for Hokkaido, the American geologist Benjamin Smith Lyman carried out a three-year geological survey: first to map the region's existing coal seams, and then to research their depth and expanse. Following the findings of his research, the Horonai Coal Mine, the largest in Hokkaido, was opened in 1879 in Mikasa.

In 1889 when the Japanese government decided to sell the Miike Coal Mine, both Mitsui and Mitsubishi were interested, and the government accepted the highest offer from Mitsui. Mitsubishi took over the Takashima Coal Mine. From that time, most coal mines were managed by the big three Zaibatsu (Japanese conglomerates usually centred on a major trading house) of Mitsui, Mitsubishi, and Sumitomo, and some were also owned by various mid-sized Zaibatsu. In the latter half of the Meiji Period, coal production was rapidly increased. However, many serious gas disasters occurred during this time. For example, in 1899, the Toyokuni Coal Mine disaster happened, resulting in over 200 deaths. In the first half of the Taisho Period (1912–1919), because of the economic boom, the number of coal mines was increased and the number of coal companies expanded from 1914 to 1919. However, the economic depression of the 1920s resulted in the coal companies' bankruptcies and mergers. Consequently, only the big coal companies could survive. It was a tumultuous era for the coal mining industry.

In the Showa period, from 1926 onwards, the government expanded the domestic coal industries to Karafuto, Sakhalin, Manchuria, and North China (now known as Shandong province in China) in order to extract high-quality coal. As a result of the growth of the munitions industry, the Japanese imperial government was eager to increase coal production. As there was a shortage of materials and manpower, the government made up for this shortage by employing workers from the Korean Peninsula. Furthermore, mechanization of coal production by that time had already begun. Consequently, domestic coal production reached 60 million tons in 1940, the highest ever production figure achieved in Japan. However, from 1940 onwards, coal production decreased, and the deteriorating war situation further reduced the manpower available. As a result, the government was forced to use a large number of unskilled miners in coal mining. Finally, at the end of World War II (WWII) in 1945, coal production was reduced drastically to 20 million tons annually.

After WWII, in Japan there were fundamental socio-economic changes that also affected the coal industry. In the early post-war years, Japan experienced a great shortage of coal. The government decided to embark on the unique strategy of the "priority production method" which primarily involved prioritizing the use of materials for the coal and steel industries in order to aid economic recovery. As a result, coal production increased, and, at that time, coal became known as "black diamond." The number of coal mine companies was about 400 at the end of WWII, but there were over 900 by 1950. From 1947 to 1949, the coal industry was state controlled. Figure 3.1 shows a map of coalfields in Japan. Since then, because of the Korean War, 1950–1953, the coal industry faced a special procurement boom. At the time, coal production recovered to 45 million tons annually. After the Korean War, many small coal mines in Japan were

Fig. 3.1 Major coalfields and major cities in Japan. (© The Japan Institute for Labour Policy and Training. Cited: Shimazaki, N. (2015a), Fig. 2)

closed one after another. At the same time, coal prices rapidly rose compared to oil, and the primary energy source changed from a "coal first and oil second" scenario to an "oil first and coal second" situation. It was an energy revolution. As a result, in 1955, the Act of Temporary Measures concerning Rationalization of Coal Mining was enacted. It meant that the decline of the coal industry began, and the development of mining mechanization took off.

Next, from 1959 to 1962, the labour union movement in Japan was at its strongest point of agitation. This was due to the rationalization of mines, which had resulted in a large-scale reduction of miners. The Mitsui Miike Strike, which occurred from 1959 to 1960, was a major incident in this era. On 29 August 1959, the Mitsui Company laid off 4580 workers and, on 2 December 1959, a further 1492 workers. However, when 1278 workers rejected the order on 11 December 1959, the company's management declared the workers' action was illegal because it disrupted production. These workers were formally discharged from the company. Their trade union, the Miike Labour Union, protested and mobilized the workers to carry out extended strikes. As a result of the company management's response, in March 1960, the union broke up into two distinct unions, the Miike Labour Union and the Miike New Labour Union. On 11 November 1960, the strike finally ended with the defeat of the left-wing Miike Labour Union. The latter half of this chapter describes the strike in detail. This strike was a matter of national concern. However, after the union's defeat, the labour movement quickly declined in Japan. At the same time, the Tanro, which was the umbrella organization of coal mine unions, demanded the nationalization of the coal industry and mobilized for a national struggle against the state. However, nationalization was not realized. From the latter half of the 1950s to 1962, the number of coal companies drastically decreased. Consequently, by 1962, there were less than 300 coal mines in Japan.

From 1963 to 1968, Japan achieved the greatest economic growth in a period commonly known as the income-doubling decade. However, the Japanese coal industry was forced to transform itself in order to survive. It was clear that coal could not compete with oil. Therefore, the government embarked on its "Scrap-and-Build Policy," which aimed to lower production costs and improved the industry's ability to compete by making distinctions between mines which should be expanded through government investment and mines to be left as they were or marked for early closure. From the introduction of the First Coal Policy in 1963 onwards, the coal industry was at the mercy of a series of nine major changes in coal mining policy over a 40-year period up to 2002. As a result of the "Scrap-and-Build Policy," by 1968, the number of Japanese domestic coal mines fell under 100. As rapid economic growth ended between 1969 and 1972, the Fourth Coal Policy was implemented. During this period, many coal companies, which were not planned to be scrapped, transformed themselves into other industries. As a result, more mines than expected were closed. However, at this time, because other Japanese industries were experiencing economic boom, the transition of discharged workers was reasonably smooth as the booming industries absorbed the retrenched workers.

Unfortunately, in 1973 the global oil crisis occurred as a result of the Arab-Israeli conflict and the resulting embargo on oil by the Organization of Petroleum Exporting Countries (OPEC) countries. The Japanese economy went into decline for the first time in the post-war era after experiencing double-digit income growth in the 1960s (income-doubling era). The 1973 Oil Crisis (later followed by the 1979 mini oil crisis, when the Shah of Iran was disposed by the religious clergy headed by the Ayatollah) meant a sudden economic recovery for the coal industry, as coal was re-evaluated as an important energy resource. However, for the companies opposed to the Fifth, Sixth, and Seventh Coal Policies, most of them were in a weak economic position and slowly became obsolete, and as a result, only the major Japanese coal mining companies could survive. In 1985, the Plaza Accord was signed in New York City and the value of the yen rose. Due to the strong yen, differences between domestic and foreign prices were accentuated, and finally the steel industry could not continue to buy domestic coal at high prices. Consequently, in 1997, the Mitsui Miike Coal Company closed, and in 1999, the government declared the conclusive end of the coal industry. After this policy decision, the government changed its coal mining policy to encouraging coal mining companies to export high-tech mining technologies overseas. At present, only one coal company, the Kushiro Coal Mine, continues implementing this policy. The chapter contributors visited the Kushiro Coal Mine in July 2017—details of the visit and observation studies can be found in Chap. 4 by Tai Wei Lim on post-coal mine closure gentrification.

Quantitative Analysis of the Japanese Coal Mining Industry

After WWII, the Japanese coal industry rapidly experienced periods of rise and decline. This can be confirmed by government statistics; Fig. 3.2 shows five indices. The first index is the annual coal production. In 1945, at the end of the war, coal production was 22,335,000 tons. In the post-war recovery period, production rapidly increased and as a result it reached 46,490,000 tons in 1951. Annual coal production mostly continued on an increasing trend until 1961 when it reached a peak of 55,413,000 tons. From then on, production began to decline, and from 1965, this decline rapidly accelerated. Consequently, by 1971, annual production fell to 31,728,000 tons. From the latter half of the 1970s to the first half of the 1980s, production fell only by approximately 20 million tons; however, from 1983 to 1991, the decline accelerated again by about another 10 million tons.

The second and third indices represent the numbers of productive coal mines and mine workers respectively, both showing earlier and greater-magnitude declines than coal production. In the post-war recovery period

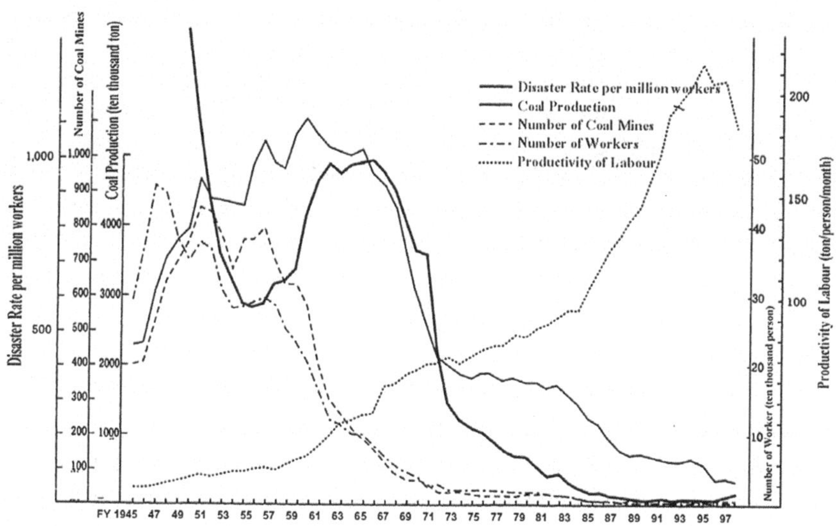

Fig. 3.2 History of coal industry in Japan: 1945–1997. (Extracted from "Coal Note 2003")

(1945–1951), the number of coal mines increased alongside coal production, resulting in 850 coal mines in 1951. Apart from a temporary improvement between 1953 and 1957, the number of productive coal mines rapidly decreased, and in 1970 productive mines numbered less than 100. Finally, by 1991, there were only a few mines that remained opened and operational. In the third index, the number of mine workers declined even earlier/faster than the number of mines. This decline had already begun in the latter half of the 1940s, and in 1949, the number of coal miners reached a record post-war high of 492,324. However, the figure rapidly declined to less than 400,000 persons in 1955. After a short period, from 1953 to 1957, when the number of miners was stable and steady, from 1957, there was yet again another rapid decline. This mirrored the decline in the number of mines. Finally, the number of working miners fell to less than 20,000 persons in the early 1970s. Clearly, the number of coal mines and miners had already begun to decline in a period when coal production was still increasing. Of course, this means that productivity per worker had increased in that same period.

Productivity kept rising through the post-war period, and in the 1960s, it shows rapid growth, from 14.2 tons per person per month in 1960 to 34.1 tons per person in 1969. This was the outcome of rationalization and modernization of coal mine production under the government's "Scrap-and-Build Policy." Unfortunately, in the fifth index, the disaster rate did not decline along with the increase in coal productivity. The disaster rate was critically high soon after the war, but it rapidly improved from 1950 to 1959, from 1397 cases per 1 million persons in 1950 down to 640 cases in 1959. The number of disasters grew worse through the 1960s because mining techniques and disaster prevention could not keep up with the deepening of coal seams. Finally, there was improvement in safety records yet again in the 1970s, but in the decade after that, there were still a number of disasters throughout the 1980s.

History of Mine Closures in the Post-War Period

In 1955, the Act of Temporary Measures concerning Rationalization of Coal Mining was enacted. Figure 3.3 shows the resulting number of coal mine closures and discharged coal mine workers from 1956 onwards. The number of closures increased sharply from 1959, reaching a peak in 1963 when the First Coal Policy was introduced in Japan, and almost 150 coal mines were closed and 18,674 miners were discharged in that era. From

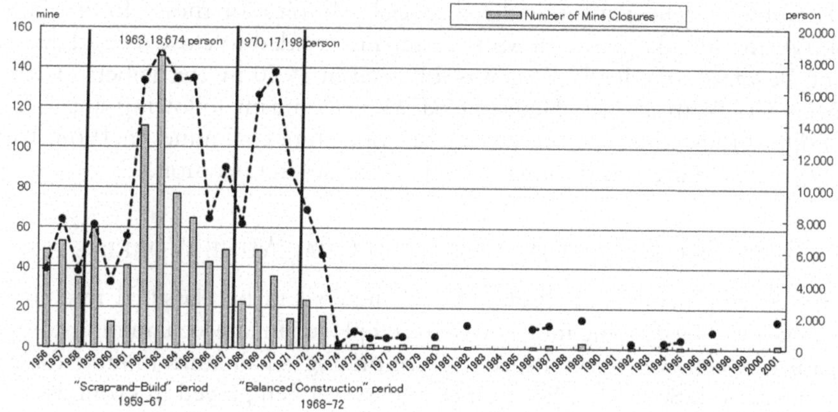

Fig. 3.3 Mine closures and the number of discharged miners. (Extracted and compiled from Coal Energy Centre 2003)

1962 to 1965, there were approximately 17,000 discharged miners annually on the average. During this period, closed mines were mainly owned by small- and mid-sized companies in the Kyushu area. Though Kyushu's mines belonged to an early phase of coal mining development in Japan, these mines were already in a final phase of coal production by the end of the 1960s.

By the end of the 1960s, the number of closures had decreased annually to about 50 mines, and in the early 1970s to about 20 annually. However, the number of discharged miners did not show the same trend; for instance, there was another peak of 17,198 discharged miners in 1970. This was a result of large-scale coal mine closures. The high number of disclosures and discharged miners during the period 1959–1967 was caused by the "Scrap-and-Build Policy," which aimed to lower production costs and improve the industry's ability to compete. During this time, a distinction was made between mines which should be expanded through government investments and mines to be left as they were or marked for early closure.

From 1968 onwards, the situation had changed so that even expanding mines or those maintaining their status quo should be closed if the coal mine company decided to move its business away from the declining coal mining industry to other growth industries. Therefore many large coal companies which had may collieries decided to liquidate their companies.

The government institutionalized special subsidies for these closures. In 1964, the Meijikogyo Coal Mine Company decided to close its coal mining business completely. This was followed in 1970 by the Yubetsu Coal Mine Company and the Haboro Coal Mine Company following suit. As a result of these closures, there were 142 operating coal mines in 1968. By 1971, this figure had declined to only 70 mines in operation.

Re-employment of Displaced Coal Mine Workers

The re-employment of displaced coal miners can be classified into five categories.[1] First, some miners were re-employed in new coal mining companies which were established for continuing small-scale mining for a short period. Second, other former miners were employed in group business companies or companies affiliated to the original coal mining companies that shut down their mines. This was a result of the Third Coal Policy, introduced in 1967, in which the government encouraged coal companies to diversify their business. A third source of employment was with companies which had been encouraged to relocate in former coal-producing areas.

In these three categories it meant that miners who were re-employed did not need to relocate. Indeed, most displaced miners preferred to stay in the same area after changing jobs. However, the possibility of re-employment for the miners in these three categories depended on the particular arrangements made by each of these coal mining companies that were shutting down their coalfields. For instance, in the case of the Joban Coal Mine's closure in Iwaki City of Fukushima Prefecture in 1971, where 4700 workers were discharged, there were 21 group business companies, called "All Joban." The *Joban Hawaiian Center*, now "Spa Resort Hawaiians," was the only successful example of a coal company transforming from the coal mining industry (heavy industries) to the tourist sector (service and retail sector) in Japan. The mining conditions of the Joban Coal Mine Company had been very harsh for the workers, in that 40 tons of hot water at temperature of about 50 °C had to be pumped out in order to extract 1 ton of coal. The Joban Company had been struggling with this problem. In the mid-1960s, the Joban Company President Yutaka Nakamura, as the founder of the restoration of Joban, proposed the idea of establishing the Hawaiian Center (a spa resort) to reuse the hot water

[1] Shimazaki, N. 2015, 'Support for Workers Displaced in the Decline of the Japanese Coal Industry: Formal and Informal Support,' "Japan Labor Review" Vol. 12, Number 2, 6–27.

from the pit's interior for hot-spring treatment. At first, due to the unconventional nature of this idea, there was unsurprisingly tremendous opposition. However, he overcame this, and finally the resort opened in January 1966. The Joban Company transferred many miners from the mines to enter the service trade at the resort and hired their family members as new staff personnel as well, for instance, the daughters of miners as Hula dancers. The movie *Hula Girl* (2006) depicted the experiences of these young women, and their hesitation and preparation to rejuvenate the town's economy. In 2016 the Spa Resort Hawaiians passed the 50th-year anniversary of its founding. At present, it continues to be a central location for leisure in the Tohoku region.

The fourth category of re-employment involved many miners moving to industrial companies located away from the original coalfield areas in major cities like Tokyo and Osaka. This became the major method of re-employment for displaced miners. During the period of economic growth in the 1960s, Japan's manufacturing industry and construction industry demanded a large number of heavy manual workers who were willing to work in a shift system. In general, coal mining work involved shifts and exhausting labour underground. Therefore, discharged coal miners provided a desirable source of labour. Some of these companies employed over 100 workers at a time. In such cases, former miners moved together and entered a new company as a group. As a result, most displaced coal mine workers sought new jobs in other industries and moved to urban regions with their families.

The final category of re-employment of former coal miners was with other coal mine companies. In the 1960s there was a chronic lack of labour in the "build mines," especially in Hokkaido (Ichihara 1997, 371). Build mine companies had to ensure that there was a continued supply of coal, and therefore further "labour-related mine closures" were avoided. The distribution of workers into these five categories depended on both the macroeconomic and socio-economic situation of each coal mine and coalfield closure, and the individual characteristics of each miner, in criteria such as age, qualification, and level of skills.

It is important to state that measures for miners' potential re-employment were only applicable to those miners in large-scale coal mine companies, not for those working in small- and mid-sized companies that operated mainly in the Kyushu area. During the period from 1962 to 1965, as described in the preceding section, there were approximately 17,000 miners discharged annually in the Kyushu area. These miners faced a completely different and more severe situation than those of large-scale companies. However, it is difficult to know the details of their re-employment because the situation of closure and the company size varied. The hardship of discharged miners depended on whether the coal company received the state subsidy for mine closure, which covered the company's debt, but did not cover workers' unpaid wages prior to closure and discharge allowance. In addition, in some cases, the company had not paid social security premiums, and so discharged miners did not receive the Black Notebook, which is the Unemployment Insurance System for Displaced Miners, and could not find new jobs. As a result, as soon as they were discharged, they fell into a life of poverty. After that they tended to become dependent on Supplemental Security Income. Clearly, there was a hierarchical structure in the Japanese coal industry based on company size, and discharged miners' re-employment prospects reflected this dual structure depending on whether they worked for large-scale company or small- to mid-sized companies in Kyushu. Fifty years later, in the present, the cumulative effects of this dualistic structure still affect several generations of residents socio-economically in the former coal mining areas of Kyushu.

Features of the Coal Mining Industry in Japan

The Japanese coal mining industry, while having certain features in common with other countries, also had its own unique country-specific characteristics/features. The first feature is the tailoring of different mining methodologies for different coal mine situations based on the coal seams' unique condition at each coalfield. In the case of Japan, the depth of mining areas made the mechanization process of extracting coal resources more difficult. Consequently, the Japanese coal industry did not succeed in coal mining modernization, and remained as a labour-intensive industry. The second feature is the characteristics of coal mining companies. The Japanese coal mining industry was mainly managed by major Japanese Zaibatsu companies, like Mitsui, Mitsubishi, and Sumitomo. After WWII, these companies developed productive mines outside Japan, for example, Australian and Canadian open-pit mines, and moved into other industries (like chemicals and heavy manufacturing industries) in the immediate post-war period. Consequently, the coal industry gradually lost out to the new sunrise growth core businesses/industries of each Zaibatsu conglomerate.

The third point is that Japanese coal companies offered a generous welfare system for its employees, for example, company housing. Therefore it could be said that miners and their families were an integral part of the company's production system. Fourth, as a result of the above situation, the position and self-esteem of mine workers were correspondingly low. It should also be noted that each company had its own union, which inevitably weakened the negotiating position of workers. Therefore, any labour movement had limited impact on the company management. It could be observed that union weaknesses resulted in a lack of safety awareness and a high rate of accidents. In addition to these four points, Williams' comparative study of Japan and New South Wales recognized two further characteristics of Japanese mine closures.[2] One was that, when facing mine closures, most Japanese coal companies immediately evacuated from the coal-producing regions, and as a result, discharged miners and their families lost their housing privileges and the way of life that they were accustomed to in coal mining areas. Consequently, most miners had to leave the coal mining regions, which rapidly declined after the mass exodus. The second characteristic concerned the regeneration of former coalfields. Williams indicated that, in Japan, the policy of regeneration was decided only by the government and companies and did not involve coal miners and trade unions. This is the most idiosyncratic and distinguishable feature of the regeneration process in Japan.

In considering the mine closures and re-employment of discharged miners, Williams' observations were accurate. However, there are other important characteristics which have to be recognized in analysing the last phase of the Japanese coal industry in the 1990s. This involves a classification of coal mining regions. The Mitsui Miike Coal Mine, in Omuta City, which is the focus of this chapter, and the Taiheiyo Coal Mine, in Kushiro City, which is described in Chap. 4 by Tai Wei Lim, were not applicable to Williams' first point. These mines can be classified as *city mines*, not just as coalfield regions. It means that the coal industry was of course one of the core industries in this area, but there were other important industries such as the chemical industry in Omuta City and the paper industry and the marine product processing industry in Kushiro City. In the case of a city mine facing closure, there were more new jobs available than in other

[2] Williams, C., 2010, 'The Rise and Fall of Coal in Comparative Perspective: Thoughts on South Wales and Japan,' Paper for "Symposium: Current Situation in Former Coalfields: Joban, Sorachi and Wales," at Waseda University.

areas, and the probability of discharged miners migrating to other regions for re-employment was relatively low. Therefore, local governments had to play a leading role in creating and developing new industries.

Overview of Miike's History

The first part of this chapter mentioned how the discovery of coal in Japan in 1469 described as "combustible stone" took place at Miike Inariyama, now called Omuta City. In 1873, during the early Meiji Period, the Miike Coal Mine was first established and further developed by the government. In 1889 the government decided to sell the mine to Mitsui Kumi. In the same year, Mitsui Kumi established the Mitsui Mining Company for managing various coal mines, including Miike. In 1911 the Mitsui Mining Company became a stock company. Following this, Mitsui developed as one of the major five Zaibatsu of Japan. In focusing on the Mitsui Miike Coal Mine, it is important to consider five important characteristics. First, it should be emphasized that the establishment of the Miike Coal Mine was the beginning of Mitsui's business dominance of the whole region. Second, Miike had the largest coal seams in Japan and its coal was the highest quality kind used for coking. These two characteristics resulted in Mitsui's predominant position in the Japanese coal mining industry.

In contrast, there were three important negative aspects of Mitsui's business. One was the use of prisoners in mining. In 1883, under government management, *Miike Shujikan* prison was established in Omuta City and prisoners who had a long prison sentence were forced to labour in the mines. After the prison management changed from the government to Mitsui, use of prison labour continued because of Mitsui's management policy of efficiency. This resulted in the number of prisoners working in Mitsui's mines in 1888 reaching 2100 of the total workforce of 3300. These prison labourers suffered from severe mining conditions, which resulted in many injuries and deaths. Indeed, during the government management period, there were several prisoners' rebellions. Under government management, the use of prisoners in coal mining was found not only in Miike but also in Takashima in Kyushu, and in Horonai and Kushiro in Hokkaido, where large prisons were established. However, in these coalfields, the modernization and mechanization of mining, after ownership was transferred from the government to private companies, brought forward the abolition of prisoner labour. For instance, Takashima Mitsubishi in Kyushu ceased this practice in 1889. Finally, in 1930 Mitsui decided to stop using prisoner labour. Up until this time, through its powerful business influence, Mitsui had been able to pressurize the government to maintain the prison in the Miike area, providing a convenient source of labour. Consequently, Mitsui's decision to stop using forced prison labour was much later than other coal mining companies.

The second negative aspect of Mitsui's businesses was the management of the strike in 1959–1960. During WWII, Miike continued its production under a government order. After the war, Miike employed many new miners. Some were Japanese overseas returnees who had been working in Japan's colonial territories in China, and others were Omuta residents who lost their business as self-employed workers during the war. The final group of new employees were demobilized soldiers. In 1946 the Miike Labour Union was established with 20,000 members. At first, this union embraced cooperation with the company. However, it shifted to a principle of resisting the company's policy of rationalization. This shift in union attitudes towards management was a result of the company's decisions to discharge many workers and reduce wages, as a result of persistent frequency of mine disasters and accidents. From 1952, the union began to go on strike.

In 1953 the Mitsui Company proposed a rationalization policy which included discharging 5738 workers from all mines, which was about 12% of its coal mining workforce. As soon as this proposal was made, all Mitsui miners' unions went on strike.[3] At Miike itself, the company proposed the discharge of 2700 workers. The Miike Labour Union went on strike, and finally the company dropped this plan. The union took pride in this first victory, and from this landmark strike onwards, it increased its influence and became the strongest union in all of Japan. This strike came to be known as the "113 day strike without one specific hero." After this strike, the Japanese government started formulating an official post-war coal policy. In 1959 the largest strike in Japan occurred. This was the *Miike Strike*. In August 1959, the company declared it was going to discharge 4580 workers from all mines. The Miike Labour Union opposed it. In December the company targeted 1278 union activists who were regarded as the culprits for disturbing production for discharge.

[3] Taniai, K. 2017, 'Miike tanko no rekishi,' in Kansai Daigaku "Tanko no kioku to kansai," 50–54 (in Japanese).

In January 1960 the company locked out the miners from the colliery and the union went on strike, which lasted for a further ten months. One of the major characteristics of this strike was the confrontation between the labour unions. In March 1960, the Miike New Labour Union was formed with the company's endorsement.[4] The members of the New Union restarted mining work, which meant breaking up strikes. The battle between the Miike Labour Union and the New Union became serious and long term. Finally after a series of mishaps, on 29 March 1960 it resulted in the death of a Miike Labour Union member. In April 1960 the strike became nationwide and the Tanro, which was a ruling body of coal miners' unions, took the leading position in the strike. Many labour unions, not only coal miners' unions but also other industry workers' unions, were marshalled and mobilized to support the Miike Strike and helped fund the Miike workers. This strike was termed "All Capital versus All Labour." This turn of events attracted the attention of the whole nation. Since the Miike Labour Union seized the hopper which was a storage facility for processed coal, the company could not take possession of it. Therefore, the company filed a provisional disposition to make the union's access to the hopper and interference with the company's activities unjustifiable. The court ordered the provision to be carried out. However, the Miike Labour Union fought back. Finally, on July 17 the "Hopper battle" occurred. The union picketed the hopper with 20,000 workers and a police force of 10,000 opposed them. Both sides were willing to die in defending their cause.

Facing this tense situation, the chairperson of the Central Labour Relations Commission started to mediate between the two sides, and finally a blood bath was avoided. On 10 August 1960 the chairperson mediated the proposal that the company would drop the discharge of designated workers and, instead, 1278 workers would be made voluntarily redundant. The Tanro accepted this proposal, and it persuaded the Miike Labour Union to do likewise. Finally, on November 1 the Miike Labour Union stopped the strike. The consequence of this strike was that of the total number of 14,805 workers, the number of union members drastically declined from 11,711 before the strike to 4834 by the end of 1961. In contrast, the number of New Union members was 6112 at that time. After the strike, new employees were forced to enter the New Union and the Miike Labour Union continued to decline, so at the Miike closure in 1997, its members totalled only 274.

[4] Chimoto, S. 2017, 'Mitsui sogi ni okeru zenro ni yoru shien,' in Kansai Daigaku "Tanko no kioku to kansai," 87–101 (in Japanese).

The third negative aspect of Mitsui's business was two severe disasters. The first disaster was the coal dust explosion in 1963. At 3:12 (1512 hours) in the afternoon of 9 November 1963, three years after the Miike Strike, coal dust exploded underground in the Mikawa Colliery of Miike. This disaster was the worst in Japan since WWII because of the loss of life and the number of injuries. The death toll from the explosion itself was 20. However, by 3 December 1963, there were a further 458 deaths from carbon monoxide (CO) poisoning, and one month after the explosion there were still 839 patients who were suffering from symptoms of anthracemia. This situation was due primarily to the doctors' and the company's lack of specific knowledge about anthracemia. These patients continued to suffer from the after-effects of poisoning. At present there are still many hospital inpatients. The patients look normal in appearance, but their behaviour is abnormal and shows evidence of child disorders, such as communication difficulties and frequent fits. Therefore, families have to keep an eye on them all day long. Consequently, not only the victims but also their families, especially their wives, have been facing indescribable distress over a period of 50 years.

One main reason for their distress after the explosion was the problem of identifying the cause of the accident. At first, the authorities determined that the accident was beyond the company's control. However, the union continued to consider that the cause of the accident was due to company negligence. After a prolonged trial, finally in 1998, 35 years after the accident, the Supreme Court made the final decision that this disaster was caused by exploding coal dust which had piled up because of insufficient cleaning. However, for many people this decision came too late. It is important to consider the background of this severe accident. After the Miike Strike, both the Miike Union and the Miike New Union became weak, and the company developed its rationalization policy. Indeed, before the accident, the company decided on a large reduction of the workforce, which resulted in a lack of safe administration, and an ineffective safety commission at each workplace. Therefore, many small accidents occurred one after another. Unfortunately, this disaster happened just two days before November 11 when labour management negotiations were held on this issue.

Since the decision regarding the cause of the disaster was delayed, the bereaved families and the patients were forced to fight for major compensation claims. The trials continued for a long time. During the claim process, they faced many difficult problems; for instance, a schism in the plaintiff between the patients and the bereaved families, and the conflicts between the plaintiff and the union because of differences of opinions. It was the wives who acted on behalf of the patients, and they were at the forefront of bargaining against the government. As a result, they fought to secure many measures, for instance, the 1966 Act on Temporary Measures for CO poisoning workers. The key figure was Mrs. Keiko Matsuo, who was 32 years old at the time of the disaster and whose husband was affected by CO poisoning and died in 1994. The documentary film *Echoes from the Miike Mine* (2006), which is the focus of Chap. 6, was focused on her long struggle. She is now 86 years old and continues the movement. The wives of the victims brought action for damages against the Mitsui Company in 1972. Finally, in 1993 the Supreme Court accepted the culpability of the Mitsui Company and certified all claimants as suffering from the after-effects of CO poisoning. However, the compensation money was not accepted by the patients' families.

After the coal dust explosion in 1963, the Miike Coal Mine turned to increasing the production efficiency. As a result, in 1970, it recorded 6,570,000 tons, which was the largest annual production of coal in Miike. It meant the daily production was over 20,000 tons. However, there was another tragedy. On 18 January 1984 the second severe disaster occurred in the Ariake Colliery of Miike. This disaster was a mine fire and resulted in 83 deaths and 13 patients suffering from symptoms of anthracemia. The Ariake Colliery was the last colliery to be built in Japan, which was developed in 1976 under the "Scrap-and-Build Policy." It was the most modernized mine with the most advanced computerized monitoring control system operating 24 hours a day. Unfortunately, two ionization smoke detectors were out of action at the time of the fire (Fig. 3.4).

Overview of the Miike Mine Closure (1997) and Omuta City

The Mitsui Miike Coal Mine was the largest coal mining company in Japan. However, it also closed down by the end of the twentieth century, as did most other coal mines in Japan. Facing the coal industry's decline in 1980s, Miike and Omuta City sought to transform the perception of the region from "coal mining town" to "town with coal mining" (i.e. an industrial area that included coal mining), and so avoiding the negative

Fig. 3.4 The flag of the Miike Labour Union. (© The Omuta City Library)

label of "town where there used to be coal mining."[5] This section of the chapter addresses the question of whether this transformation was successful. This chapter focuses specifically on three aspects of the process of decline which Miike and Omuta City experienced. First is the issue of Miike's presence in Omuta City. Second is Miike's rationalization programme from 1987 onwards. Third is Miike's closure in 1997 and the regeneration of Omuta City, in the aftermath, over a 20-year period.

Miike's Presence in Omuta City

Since the 1970s Mitsui has had a major presence in Japan on a national scale. Also, while coal mining production nationally declined quite rapidly in the 1970s, as described in the previous section of this chapter, the decline of Mitsui Miike's mining in Omuta City was less drastic. Though from 1970 to 1980, Miike's annual coal production had decreased from 6,570,000 tons to 5,340,000 tons, its market share had increased from 17.1% in 1970 to 23.2% in 1973 and to 29.5% in 1980. Miike and Omuta City became a core coal-producing area.[6] It can be said that

[5] The Miike Coal Mine Company's paper "*KURODAIYA*" 15 October 1990 (in Japanese).

[6] Kyushu Economic Research Association 1987, "Research Report: Coal Mining Industry's Effect on Omuta City" (in Japanese).

Omuta City's industry originally relied on the Mitsui's coal mining and coal-chemical industries, and Omuta City became Mitsui's company town. Table 3.1 shows the number of companies and employees of the Mitsui Coal Mine Group and its affiliated companies in 1986.[7] There were nine companies that belonged to the Mitsui Coal Mine Group. Their business covered all the coal mining industry's activities, such as transportation, machinery, electricity, sales, and housing. The total number of employees of these companies was over 7000. Furthermore, there were 179 subcontract companies and affiliated companies, mainly in the areas of coal mining, construction, commerce, and transportation. The number of employees of these companies was up to 7793, which was more than the Mitsui Group.

Table 3.1 Affiliated companies of the Mitsui Coal Mine (1986)

	Number of companies	*Number of employees*
Mitsui Coal Mine Group	9	7183
Mitsui Coal Mine Co.	1	4802
Miike Port Co.	1	532
Mitsui Miike Machinery Co.	1	811
Mitsui Construction Co.	1	45
Ariake Machinery Co.	1	479
Ariake Electricity Co.	1	222
Miike Business Co.	1	89
Sanko Ariake Housing Co.	1	9
Mitsui Greenland Co.	1	194
Subcontract companies and affiliated companies group	179	7793
Omuta Industries Association	10	877
Ariake Industries Association	11	744
Sanko Construction Association	50	3035
Sanko Commerce Association	73	1950
Miike Port Affiliated Companies	25	964
Others	10	223
Total	188	14,976

Source: Kyushu Economic Research Association (1987), Table 1-3

[7] Kyushu Economic Research Association 1987, "Research Report: Coal Mining Industry's Effect on Omuta City" (in Japanese).

Table 3.2 Number of employees of major companies and businesses in Omuta City (1960–1986) as given on April 1 of each year

	1960	*1965*	*1970*	*1975*	*1980*	*1986*
1. Mitsui Miike Coal Mine	12,822	10,444	7822	6280	5242	4288
2. Denka, Omuta Factory	1582	1567	1156	654	607	650
3. Mitsui Chemicals, Omuta Works	7317	5679	3216	2326	1710	599
4. Mitsui Coke Engineering Omuta Factory	–	24	564	600	416	199
5. Mitsui Aluminum, Omuta Works	–	–	202	818	881	703
6. Mitsui Metal Miike, Miike Smelting	2333	2108	1670	941	586	271
7. Mitsui Miike Machinery Co.	1694	1930	2036	1213	1273	672
8. Miike Port Co.	1347	1155	889	674	594	434
9. Kyushu Electric Power, Port Plant	817	555	293	176	133	136
10. Miike Gosei, Miike Factory	952	–	–	–	–	–
Total	28,864	23,462	17,848	13,682	11,442	7952
Reference						
Concerning manufacturing (except 1, 8, and 9): A	13,878	11,308	8844	6552	5473	3094
Total number of employees in manufacturing in Omuta City (Industrial statistics): B	18,374	16,902	15,056	13,283	11,431	9896[a]
A/B (%)	75.5	66.9	58.7	49.3	47.9	38.1[a]

Source: Kyushu Economic Research Association (1987), Table 1-1

[a]1983

Table 3.2 shows the number of employees of ten major companies and businesses in Omuta City, including the Mitsui Miike Coal Mine and its affiliated companies.[8] All of their businesses were heavily related to the coal industry, such as producing chemicals, coke, aluminium, and electricity. In 1960 the total number of employees of these companies peaked at 28,864 persons. By 1986 the number had rapidly declined to about 8000. The Reference section of this table shows the total number of manufacturing workers employed in Omuta City, and it shows the percentage of employees working in the manufacturing sector of these major companies. In 1960 75% of manufacturing workers in Omuta were employed in these coal-related companies.

[8] Kyushu Economic Research Association 1987, "Research Report: Coal Mining Industry's Effect on Omuta City" (in Japanese).

However, this figure declined after 1960, finally falling to below 40% by 1983. Of course, the coal industry had a major impact on the Omuta economy. For instance, in 1983 the gross output of the mining industry was 45 billion Japanese yen, which was 14.8% of the gross output of Omuta City, and the number of employees was 7594 persons, 10% of the total.[9] It is clear from the above statistics that the Mitsui Miike Coal Mine played a key role in the Omuta economy and had either directly or indirectly impacted it for a long time. In line with the general decline of the coal industry, Miike's influence also tended to decline. However, Miike's presence in Omuta City was still quite substantial in 1987 when the Eighth Coal Policy was initiated.

The Eighth Coal Policy and Miike's Rationalization Programme

The Eighth Coal Policy declared the government's decision to change its basic stance from the rationalization of the Japanese coal industry (which had aimed to maintain the industry's operational status) to a policy to bring about the structural adjustment of the Japanese coal industry with a long-term aim of final closure. This change was first proposed by the Maekawa Report[10] in 1986, which clearly addressed the coal industry's future. That is, the coal industry had to carry out structural reforms to bring about a significant reduction in domestic coal production, while paying careful attention to the severe economic impact on coal-producing regions. This would also lead to the expansion of overseas coal importation. It was the first time that the Japanese government specified clearly that coal production had to be downsized through a structural adjustment of the coal industry.

As a result, the Eighth Coal Policy subsequently prescribed that, in the next five years after the Maekawa Report, the production of coking coal and fuel coal for Japanese industries in general had to be cut to zero, and domestic coal production had to be reduced annually by 10 million tons

[9] Kyushu Economic Research Association 1987, "Research Report: Coal Mining Industry's Effect on Omuta City" (in Japanese).

[10] Mr. Haruo Maekawa, who was the governor of the Bank of Japan, doubled as a principal of the prime minister's private advisory panel. The Maekawa Report was made by this panel, and it declared the basic principle of future Japanese economic policy and had a great impact on politics and industry.

to produce fuel coal for domestic power generation only. This policy meant the final end of the coal industry in Japan. It continued right up to 1991, and thereafter, the Post Eighth Coal Policy was implemented in 1992 and completely executed in 2002.[11] According to this Eighth Policy, Miike had to revise its coal-producing system and plan in order to meet the coal demand. Miike's coal was the highest quality used for coking, and so when the domestic coking coal market closed, it could be said that this policy indirectly determined that Miike's closure in the near future would be unavoidable. Furthermore, it would have to follow a series of incremental steps to reach that end. Subsequently, Miike began the process of closure along with its rationalization policy.

The Mitsui Miike Coal Mine had to undertake a large-scale rationalization programme from 1987.[12] It consisted of downsizing coal production and lowering coal-producing costs. The former included personnel reduction and the latter included the mechanization of coal exploitation and revision of the coal-producing system. First, concerning the downsizing of coal production, Miike planned a gradual reduction from 4,530,000 tons in 1985 to 3,500,000 tons in 1987, 3,000,000 tons in 1988, and finally 2,500,000 tons in 1989. In only five years, its production decreased by 55% from 1985. To bring about this reduction the company solicited voluntary resignation on three occasions, in 1987, 1988, and 1989. The result was the accepted retirement of 1803 miners: 487 in 1987, 698 in 1988, and 618 in 1989. One important characteristic of the workers who accepted voluntary retirement was their young age, especially for miners. Figure 3.5 shows their age distribution. Two-hundred and forty workers were aged 20–29 (19%), and 487 were 30–39 (39%), which means a total of 727 (58%) were under 40 years old. In general, these young discharged miners were able to find new jobs easily, mainly in the local regions of Omuta and Arao.[13]

Following rationalization from 1987 to 1989, the number of workers decreased from 3219 in the end of fiscal year 1986 to 1481 in the end of fiscal year 1989 (Fig. 3.6). In the year before mine closure, the workers' age distribution at Miike was highly skewed (Fig. 3.7). The figure shows

[11] Shimazaki, 2018 'Coal policy in Japan,' Nakazawa and Shimazaki eds., "Coal Mines and 'Japanese Miracle,'" Seikyusha (in Japanese).

[12] Primary documents of the Planning Coordination Department, Omuta City, #2-1-8, 1989.

[13] Primary documents of the Planning Coordination Department, Omuta City, #2-8-4, 1990.

Fig. 3.5 Age of voluntarily retired workers between 1987 and 1989. The age is as indicated as at 31 January 1989. (Extracted and compiled from primary documents of the Planning Coordination Department, Omuta City, #2-1-8, 1989)

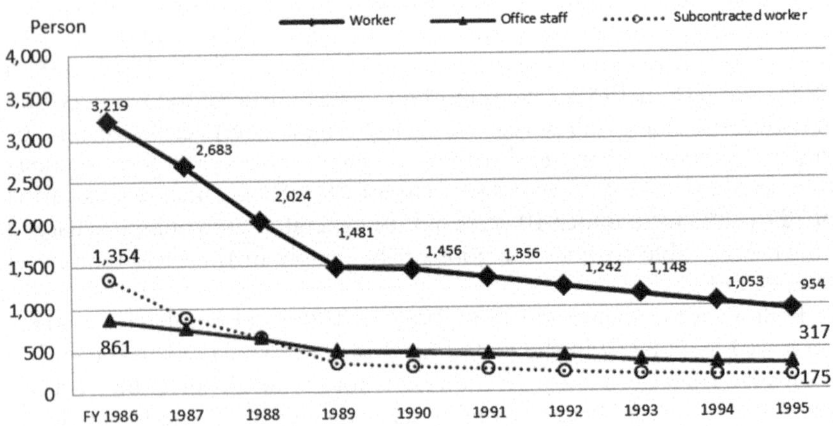

Fig. 3.6 Number of Miike workers between 1986 and 1995. (Extracted and compiled from primary documents of the Planning Coordination Department, Omuta City, #1-2-8, 1996)

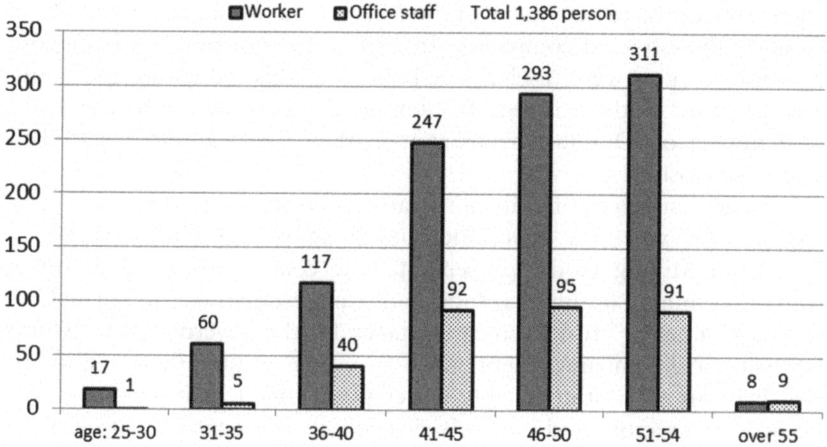

Fig. 3.7 Age of Miike workers in 1995. (Extracted and compiled from primary documents of the Planning Coordination Department, Omuta City, #1-2-8, 1996)

that miners over 41 years old numbered 851 (61%). As a result, older miners were discharged at the time of mine closure in 1997. As indicated in the previous section, Miike had reduced personnel several times, such as in 1953 and 1959. These reductions resulted in miners' strong resistance, which triggered union strikes, such as the *113-day strike* and the *Miike Strike*. However, this time strong resistance did not occur, and the reason was that the Miike Labour Union had been already sidelined, and the Miike New Labour Union, which had a collaborative relationship with the company, was a more receptive negotiation partner working with the management.

Second, in downsizing of coal-producing costs, Miike came up with two main measures. One was the rationalization of coal-producing procedures, for instance, consolidating the pitheads from three to one, implementing the mechanization of coal exploitation known as the MHP system (Mitsui high-power plant), and revising and improving the coal cleaning process. Another measure was reduction in the number of contracted workers. In Miike there used to be a clear division of labour roles for those working underground. Company employees were assigned only to coal-cutting and contracted workers to digging tunnels, installing pillars, and other peripheral jobs at the mines. The Mitsui Construction was the largest subcontracted company of the Mitsui Coal Mine, and it had five of its own subcontracted companies. In 1987 a total of 670 of Mitsui construction workers were discharged. Following this reduction, these jobs were absorbed by the existing Mitsui miners. This worker reduction had a great impact on Omuta citizens as it marked the beginning of Miike's process of closure.

It is necessary to clarify Mitsui's strategy in response to the Coal Policy. Back in 1973 when the government implemented the Fifth Coal Policy, the Mitsui Mining Company separated its coal-producing department from other mining businesses/subsidiaries and established the Mitsui Coal Mining Company.[14] It was a measure taken for the government to assume the major coal companies' debt. Coal companies wanted their coal industrial divisions to continue stably under government management, while the non-coal sector would actively diversify its businesses to support the coal sector. In 1969, Mitsubishi separated its business activities this way, and all other major coal mine companies (Mitsui, Sumitomo, Hokutan, Taiheiyo, Matsushima, and Joban) followed suit.[15]

Much later in 1986, when the Maekawa Report was published, the Mitsui Coal Mining Company was operating three coal mines, Miike in Kyushu, and Ashibetsu and Sunagawa in Hokkaido. However, at the beginning of the Eighth Coal Policy in 1987, the company closed the Sunagawa Coal Mine. At that time, the Ashibetsu Coal Mine faced the same situation as Miike, with coal demand reduction owing to the abolition of regional energy power plants. Therefore, in 1987 Ashibetsu downsized coal production from 800,000 tons to 590,000 tons, with further reductions in 1988 to 410,000 tons, in addition to rationalizing its mining areas and the transportation system.[16] Furthermore, the Ashibetsu, like Miike, discharged many miners, and voluntary resignations were 483 workers in 1987 and 171 in 1988. In this work year alone, 50 workers were temporarily transferred to other industry and 47 were forced to retire. As a result, the number of workers drastically decreased, and at its

[14] Mitsui Mining, 1990, "*Otoko tachi no seiki: Mitsui Kozan no Hyakunen*," (in Japanese) pp. 360–367.

[15] Ushijima, 2012, 'Dai yo ji sekitan seisaku to kigyo saihen,' in Sugiyama and Ushijimaeds., "Nihon sekitan sangyo no suitai: sengo Hokkaido ni okeru kigyo to chiiki," (in Japanese) Keio University Press, pp. 125–153.

[16] Mitsui Mining, 1990, "*Otoko tachi no seiki: Mitsui Kozan no Hyakunen*," (in Japanese) p. 389.

closure in 1992, there were only 356 remaining miners to be discharged.[17] After the Ashibetsu closure in 1992, the Mitsui Coal Mine consolidated its management at the Miike Coal Mine.

Miike's Closure in 1997 and the Regeneration of Omuta City

The rationalization exercise from 1987 meant that Miike's closure in the near future was unavoidable or that in effect the process of closure had already started. However, Miike still continued its coal production despite the fact that its business was shrinking. The MHP, a new coal exploitation plant, went into full-scale operation from December 1988. Daily production peaked at over 10,000 tons.[18] From 1995 onwards, the closure was targeted by the year 1997. There were two important reasons for this announcement. First, a custom duty on crude oil, included in the coal policy budget, would be reduced from 315 Japanese yen per 1 kilolitre to 215 Japanese yen in April 1997. Second, the state-determined coal price, which was the price of the power industry's purchase, would be reduced in the fiscal year 1997. Finally, in March 1996 the Miike Company unilaterally abolished the CO Agreement which was made with the three unions for the welfare and financial security of CO accidents' bereaved families and patients, which meant Miike's closure crystallized and was set in motion. The Mitsui Company assumed that this abolition was the biggest barrier to mine closure, but the three unions agreed to it. The CO accidents' patients were abandoned at the last moment.

After that, the process for closure was relatively smooth. Of course, there was an opposition campaign, but the closure was extremely straightforward. It was a "calm closure" for the workers, who generally recognized its inevitability.[19] On 17 February 1997, the company made a formal offer of complete closure to the three unions. On March 30, the Miike Coal Mine closed and 1207 miners were discharged. At that time Miike's

[17] Shimazaki, N. 2015, 'The final policy of the coalmining industry and miners' re-employment: A microdata analysis of Mitsui Ashibetsu Coalmine in Hokkaido, Japan,' (in Japanese) "The Annuals of Sociology (*Shakaigaku nenshi*)," Vol. 56, pp. 93–110.

[18] Omuta Coal Industry and Science Museum, 2014, "The History and Technology of Miike Coal Mine," (in Japanese) p. 134.

[19] Mainichi Newspaper Co., 1997, "Closure: MiTsui Miike's 124 years," (in Japanese) p. 28 and Nagayoshi, M., 2017 'Born in Coal Town,' "Memories of Coal Mine and Kansai Area: 20 years Memory after Miike Closure," (in Japanese) p. 56.

124-year history of mining finally came to an end. The unions accepted it, on May 30 the Miike Staff Union dissolved, and on the next day the Miike New Labour Union also closed down its membership. However, the Miike Labour Union did not dissolve and continued to provide support for discharged miners. When it finally dissolved on 10 April 2005, there were only 14 members. The union held a symbolic ritual ceremony and burned its flag as a sign of closure. With the unions disbanded, the discharged miners totalling 1553 faced a more severe economic situation than ever before resulting from the closure of a mine. Furthermore, they were elderly and owned homes near their place of employment, and so their prospects of moving to other locations were limited.

Soon after the closure, the Public Employment Security Office started job mediation procedures, and it mainly conducted career counselling through home visits. However, the discharged miners' re-employment was not successful. At the end of December 1997, nine months after the closure, over 530 persons were still in search of jobs. Furthermore, in December 1998, 371 persons remained unemployed, which meant the percentage of re-employment was only 59.1% of discharged miners. At that time, the job-offers-to-seekers ratio was only 0.34 in Omuta City and 0.33 in Arao City. Unemployment continued to be a very severe situation. On 30 March 2000, the time of the unemployment allowance's expiration date, around 480 persons lost their allowance and 150 persons of them became unemployed without pension. This situation was the most severe ever experienced by discharged miners. By the end of March 2001, those discharged miners under 50 years old were able to secure employment through job training, mainly in the manufacturing and construction industries. However, their new jobs fell short of their expectations in terms of payment and security of contract. The job market was relatively small, and Mitsui did not supply enough jobs. Only Omuta City government tried to develop new employment opportunities. As such, most of the re-employed miners found jobs only in Omuta, in Arao City, or in other cities in Fukuoka or Kumamoto Prefecture.[20] This scenario differed from the one experienced by the discharged miners in the rationalization process from 1987 to 1989, when they could find new jobs outside the region.[21]

[20] Employment and human Resources Development Organization of Japan, 2002, "History of Employment and human Resources Development Organization of Japan," (in Japanese) pp. 2–19.

[21] Primary documents of the Planning Coordination Department, Omuta City, #1-2-8, 1996.

As previously indicated, Miike and Omuta City sought to transform popular perceptions and the imageries of the region from a "coal mining town" to a "town with coal mining," while avoiding the negative label of a "town where there used to be coal mining." This transformation could not be declared as a complete success. At that time, one practical way to improve the image of a "town with coal mining" was through the use of high-level coal mining technology. The Japanese government implemented this during the "Post Eighth Coal Policy," when it established centres for transferring technology to developing countries such as Vietnam, Indonesia, and China. The last three coal mines of Miike, Ikeshima, and Taiheiyo started transferring technology to overseas countries as a major part of their business from the 1990s onwards.[22] Therefore, Ikeshima and Taiheiyo could transform their business activities by being entrusted with a national project. Even in 2018, the Kushiro Coal Mine, which is the successor to the Taiheiyo Coal Mine, still continues this business. On the other hand, Mitsui did not agree that Miike should transform its business in this way, so it could not continue mining after Miike's closure. Finally, the label of "town with coal mining" completely disappeared.

At present, Omuta still cannot avoid the negative label of "town where there used to be coal mining." Following Miike's closure, Omuta's population consistently declined. It peaked at 208,887 persons in 1959, 145,085 before closure in 1995, 138,629 in 2000, and 123,638 in 2010. The main sector of employment drastically shifted from secondary industries to tertiary industries. It meant that Omuta was no longer Mitsui's company town and a base for the company's business activities. However, at the same time, Omuta City could not find alternative sources of industry. Although industrial transformation is not yet a success, there are still significant signs of Omuta City's regeneration. Some activities started to improve the negative stereotypes of the Miike Coal Mine region by building on upgrading activities originally started in the 1950s and inherited by successive generations of Omuta residents and stakeholders. The final section this chapter focuses on three important activities led by key individuals who are not directly connected with the Japanese coal industry but are conscious of the important historical role of coal mining in Omuta City in the town's existence.

[22] Shimazaki, N. 2015, 'Support for Workers Displaced in the Decline of the Japanese Coal Industry: Formal and Informal Support,' "Japan Labor Review" Vol. 12, Number 2, 6–27.

The first activity is the archiving of the documents related to the history of the Miike Coal Mine. The second point is the emergence of a socio-cultural movement. The third trend is the rise of an industrial heritage movement. First, the Omuta City Library started early when it came to collecting and archiving materials related to the Miike Coal Mine, the use of prisoners in mining jobs, the Miike Strike, the Miike coal dust explosion disaster, and trials related to CO (Carbon Monoxide) poisoning amongst miners after they were admitted as hospital patients. For instance, during the Miike Strike, 1959–1960, the librarians collected almost all flyers (or leaflets) published daily by the company, the unions (specifically the Staff Union, the Miike Labour Union, the Miike New Union), and the Omuta City Government. With a total of 639 documents, the librarians made an inventory and archived the originals. At present, the inventory is one of the most valuable archives of the Japanese labour movement history.

Those curating activities were made possible at that time as the librarians were keeping sharp eyes on the chaotic situation of the Miike Strike in their region, and they felt they had to conserve its records as part of local history. Consequently, successive librarians continued to appreciate the value of this collection. A brief mention of the librarians' contributions is included here. One of these Miike librarians was Toshihide Ohara, an Omuta native born in 1951 who came to work at the library from 1985 to 1999. He was intensely interested in the Miike CO poisoning cases and Miike's closure. Through this interest, he contributed immensely to archiving and publishing the Miike documents. His action was divided into two phases: the first one revolved around Miike's closure in 1997 when Ohara was a librarian, and the second phase involved the publication of these documents as a volunteer after his retirement in 2011.

Facing Miike's closure in 1997, many newspaper and journal articles, books, and television programmes already started featuring the mine's impending closure as early as 1996. The Omuta City Library decided to collect all of them. In those days, the collection of television programmes recorded on video media was only possible at the time of the programmes' production, but with help and perseverance, the librarians successfully created a comprehensive archive of these programmes. Furthermore, at the closure of the mine, the library formally asked the company and the three unions to contribute their documents to the library. However, the company and the Miike Staff Union refused, while the Miike New Labour

Union contributed only published documents to the library. Fortunately for the curators and archivists, the Miike Labour Union donated its documents to Kyushu University, other kinds of related materials to the Omuta Coal Industry and Science Museum, and its books to the Omuta City Library at the time of its closure in 2007.[23]

One important archived set of materials found in the Omuta City Library is a collection of documents regarding the Miike CO poisoning investigations and trials. At the beginning of the trials on behalf of the CO victims and their family in 1972, the Miike CO Collegium, including plaintiff, legal defence group, doctors, and journalists, contributed all relevant documents and materials concerning the trials to the library after the trials concluded. In 1999, the year after the Supreme Court trial concluded, the Miike CO Collegium started to collect the documents and materials with the aim to publish them, and because of this, many important documents related to Miike were collected. Librarian Ohara was a member of the collegium, and with his further contributions, the library and the CO Collegium finally published the "Microfilm Corpus of the Accident of the Mitsui Miike Coal Dust Explosion," Vol. 1 in 2005 and Vol. 2 in 2007. These are highly valuable documents and interested parties must pay the equivalent of 2,800,000 Japanese yen (about $255,000) to own/access the originals. At present, many public libraries and university libraries in Japan possess them, and many social scientists use them to analyse one of the most significant social issues during the period of high economic growth in Japan.

These corpora covered the accident in great detail, for instance, the preceding issues such as the Miike Strike, trial records, medical documents of the accident, scientific documents of the accident's causation, and patients' personal documents, such as diaries, totalling over 5200 microfilm reels. Among them, Drs Shinobu Araki and Masazumi Harada's documents, Mrs. Keiko Matsuo's diaries, and the follow-up research materials of CO patients were especially valuable. Dr Shinobu Araki was an expert in studying coal dust explosions and judged the cause of accidents to be the companies' fault soon after it occurred in 1963, but his judgement of the accident's causation was rejected by the company and the government. Araki gave important evidence at the initial trial. Finally, in 1993, the court formally recognized his judgement.

[23] Ohara, T. 2015, '*Omutashiritsu toshokan ga shozo suru mike kankei shiryo to sono mokuroku ni tsuite*,' Kyushu University 30, 83–109 (in Japanese).

Araki's many detailed field notes, which were archived in the corpora, depict precise details of the situation underground in the mine. Besides Araki, Masazumi Harada was another veteran researcher and eyewitness to manmade environmental disasters in Japan's fast-growth period. He was also a key person providing medical support to sufferers of the Minamata disease in 1956, which is the worst case of health issues caused by environmental pollution in Japan. Omuta City is located near Minamata City in Kumamoto, and the Miike CO poisoning patients' trials were very much affected by the Minamata disease trials, especially in issues concerning medical causation and from the evidence presented in court.[24] Harada was also involved in the Miike CO poisoning patients' medical treatment, so he was a link that brought together the two issues' plaintiffs. In November 1996, Harada and his colleagues made a last follow-up survey of the CO patients 33 years after the accident.[25] The microfilm corpora included medical records. Furthermore, Mrs. Keiko Matsuo's personal documents consisted mainly of her diaries, personal letters, and audio-tape recording of their discussions from 1963 to present. Ohara and Matsuo had maintained an important relationship that rallied them together on this issue for many years. Therefore, Matsuo eventually gave all her documents accumulated over a long period of time to the library.

The second activity is the emergence of a socio-cultural movement in Omuta. From the 1950s onwards, the avant-garde *Kyushu Art School*, later called the Seibu Drawing School, started a drawing class in Omuta. The second president of this school, Mr. Tadashi Hataraki, who was born in 1934 and died in 1996, moved to Omuta in 1965 with his families. At this school, many young and elderly people, from not only from Omuta but also other parts of Kyushu and nationally from all of Japan, including famous social activists such as Mr. Eishin Ueno, gathered at this school and had nightly topical discussions. They were especially keen on contemporary social issues, for example, the Minamata disease, the Miike CO poisoning patients, poverty in the Chikuhō coalfield area, environmental pollution, abuse of human rights in Omuta, and nuclear energy. Around that time, Omuta and this school became centres of a socio-cultural movement known as the "Circle Movement" in Kyushu.

[24] Kumagai, H. 2012 "*Mukashi Genpatsu Ima Tanko*," Chuou Kouron Shinsha, pp. 228–260 (in Japanese).

[25] Harada, M. 1997, "Yama no Hi wa Kietemo," Nihon Hyoron Sha (in Japanese).

Mr. Tadashi Hataraki's son, Mr. Jun Hataraki, born in 1959, is an artist and poet, and he returned to Omuta from Tokyo in 1993 at the time of his father's illness and took over the school. On the day of Miike's closure on 30 March 1997, he and 20 students of his students, ranging from kindergarten children to high school students, sat down in front of the coal train station at the Miike Port and did a sketch of coal mine.[26] His aim was to get them to experience the coal mining scene of their birth place on this symbolic day emotively by heartfelt experiential learning (Fig. 3.8). Until today, he continues organizing this artistic sketching event every year and Hataraki and his circle of colleagues also inherited their elders' interests in social issues. For instance, Mr. Bunju Esaki, born in 1960 in Omuta, is a junior high school teacher, and he developed his original textbook on the topic of regional human rights abuses while pondering over the significance of his as well as of his students' birth place of Omuta.

Fig. 3.8 The artistic sketching event on the day of Miike's closure, 30 March 1997. (© Kansai University)

[26] Taniai, K. 2017, 'Miike tanko no rekishi,' in Kansai Daigaku "Tanko no kioku to kansai," 50–54 (in Japanese).

Esaki's mother is also a grammar school teacher who made efforts in disseminating human rights education.

The third activity is an industrial heritage movement. In 2001, Dr Mamoru Nagayoshi, who is of a younger stock born in 1970 in Omuta and a cultural anthropologist by training, established an non-profit organization (NPO) advocating heritage preservation and the Omuta and Arao Coal Mine town fan club with his colleagues.[27] This organization aimed to pass on the heritage gems, historical landscape, and imagination through memories (or a reconstructed mental picture) of the coal mining town of Omuta to the next generation. For this purpose, it needs to change public perceptions of the town to make these memories worth conserving. Therefore, this NPO and the Omuta Coal Industry and Science Museum (established in 1995) try to preserve the coal mining heritage, mining technologies, and memories of Omuta people's lives in their collaborative activities. Mr. Nagayoshi and Omuta City Museum curators asked Ms. Hiroko Kumagai, who is a film director, to record 32 mine sites and the biographical histories of 72 ordinary people who were related to the coal mines as an Omuta City project from 2001 to 2002. As a result, six works with documentary content were completed, which are screened in the Omuta Coal Industry and Science Museum as a "Voice Museum." Furthermore, using this content she produced the documentary film *Echoes from the Miike Mine*.

After lengthy negotiations, Miike was finally registered by the United Nations Educational, Scientific and Cultural Organization (UNESCO) in 2015 as a World Industrial Heritage, officially as the "Sites of Japan's Meiji Industrial Revolution: Iron and Steel, Shipbuilding and Coal Mining." This designated UNESCO World Heritage Site consists of 23 sections, and among the coal mine parts are four Miike heritages: Miyanohara Pit, Manda Pit (Fig. 3.9), Miike Coal Railway, and Miike Port. At present, at those sites, many volunteer guides, who were former coal miners, actively talk about their real-life experiences to visitors. To summarize, it can be said that during the last 20 years since Miike's closure, Omuta has not only un-mitigatingly succeeded in regeneration but also averted a drastic decline in some aspects. To understand Omuta's trials and tribulations, it is important to appreciate the role of these three individuals and many other stakeholders discussed here. That is, they lived their childhood in Omuta near or

[27] Nagayoshi, M. 2017, 'Tanko no machi ni umare sodatte,' in Kansai Daigaku "Tanko no kioku to kansai," 55–57 (in Japanese).

Fig. 3.9 The Manda Pit. (© Naoko Shimazaki)

in the coal mining region, and they left Omuta for out-of-town school education. After several years they returned to Omuta and found employment in their hometown. Their return and their activities were important for the Omuta region, whether in terms of revival, conservation, or awareness creation. However, it was only possible in Omuta due to its local culture and returning residents, something not found in other former coalfields, or even in urbanized city mining areas such as Kushiro and Joban. However, at present, 20 years after the mine closure, Omuta is confronting important issues of preservation of memories and tangible/intangible heritage inheritance for the next generation of Omuta residents.

In the next chapter, the historical narrative continues with Tai Wei Lim's account on gentrification. If Shimazaki's chapter focused on the "software" of coal mine closures (i.e. the individuals, families, and towns that are personally and emotionally affected by the industrial decline) and also the human-centred stories of re-employing these workers, retraining them, and counselling them, then Tai Wei Lim's next chapter is a contemporary narrative of the "hardware" (i.e. the gentrification of spaces in former coal mining towns). Together, Shimazaki's chapter with Godo's previous chapter (Chap. 2) forms the complete historical narrative of the developmental history of Japan's coal mining industry from the Meiji period to the contemporary 1990s. This chapter also bridges the chronological narrative from post-war mines closures to contemporary use of former coal mine spaces for tourism, artistic expressions, environmentally friendly spaces, and energy conservation showcases. The stories of growth, renewal, decline, and revival are weaved into the narratives of Chaps. 2, 3, and 4.

Bibliography

Chimoto, S. 2017, 'Mitsui sogi ni okeru zenro ni yoru shien,' in Kansai Daigaku "Tanko no kioku to kansai," 87–101. (in Japanese)

Employment and human Resources Development Organization of Japan, 2002, "History of Employment and human Resources Development Organization of Japan," (in Japanese) p. 2–19.

Harada, M. 1997, "Yama no Hi wa Kietemo," Nihon Hyoron Sha. (in Japanese)

Hataraki, J. 2017, 'Heizan suru tanko no sukecchi,' in Kansai Daigaku "Tanko no kioku to kansai," 129–130. (in Japanese)

Ichihara, H. 1997, "Tanko no Rodo Shakai Shi," Taga Shuppan. (in Japanese)

Kumagai, H. 2012 "Mukashi Genpatsu Ima Tanko," Chuou Kouron Shinsha, p. 228–260. (in Japanese)

Kyushu Economic Research Association 1987, "Research Report: Coal Mining Industry's Effect on Omuta City." (in Japanese)

Mitsui Mining, 1990, "Otoko tachi no seiki: Mitsui Kozan no Hyakunen," (in Japanese) pp. 360–367, 389.

Nagayoshi, M. 2017, 'Tanko no machi ni umare sodatte,' in Kansai Daigaku "Tanko no kioku to kansai," 55–57. (in Japanese)

Ohara, T. 2015, 'Omutashiritsu toshokan ga shozo suru mike kankei shiryo to sono mokuroku ni tsuite,' Kyushu University 30, 83–109. (in Japanese)

Omuta Coal Industry and Science Museum, 2014, "The History and Technology of Miike Coal Mine," (in Japanese) p. 134.

Primary documents of the Planning Coordination Department, Omuta City, #1-2-8, 1996.

Primary documents of the Planning Coordination Department, Omuta City, #2-1-8, 1989.

Primary documents of the Planning Coordination Department, Omuta City, #2-8-4, 1990.

Taniai, K. 2017, 'Miike tanko no rekishi,' in Kansai Daigaku "Tanko no kioku to kansai," 50–54. (in Japanese)

Taniai, K. 2017, 'Miike tanko rodo kumiai to kansai no rodo undo,' in Kansai Daigaku "Tanko no kioku to kansai," 76–86. (in Japanese)

The Miike Coal Mine Company's paper "KURODAIYA" October 15 1990. (in Japanese)

Shimazaki, N. 2015a, 'Support for Workers Displaced in the Decline of the Japanese Coal Industry: Formal and Informal Support,' "Japan Labor Review" Vol. 12, Number 2, 6–27.

Shimazaki, N. 2015b, 'The final policy of the coalmining industry and miners' re-employment: A microdata analysis of Mitsui Ashibetsu Coalmine in Hokkaido, Japan,' (in Japanese) "The Annuals of Sociology (Shakaigaku nenshi)," Vol. 56, p. 93–110.

Shimazaki, N., 2018 'Coal policy in Japan,' Nakazawa and Shimazaki eds., "Coal Mines and 'Japanese Miracle,'" Seikyusha. (in Japanese)

Ushijima, 2012, 'Dai yo ji sekitan seisaku to kigyo saihen,' in Sugiyama and Ushijimaeds, "Nihon sekitan sangyo no suitai: sengo Hokkaido ni okeru kigyo to chiiki," (in Japanese) Keio University Press, p. 125–153.

Williams, C., 2010, 'The Rise and Fall of Coal in Comparative Perspective: Thoughts on South Wales and Japan,' Paper for "Symposium: Current Situation in Former Coalfields: Joban, Sorachi and Wales," at Waseda University.

CHAPTER 4

Historical Development and Gentrification of Hokkaido's Former Coal Mining Areas: Case Studies of Bibai, Kushiro, Mikasa, and Yubari

Introduction

The previous chapters provided the historical framework to discuss Japan's coal mining industry. Godo gave a detailed account of the origins of the coal mining industry, in terms of both changing mindsets (e.g. adjusting to the concept of time schedules) and modern evolution from peasantry and feudal lordships in an agricultural economy to rationalized mass production in factories. Shimazaki continued with that narrative in the post-war period and discussed coal mine closures as well as policy-based coping mechanisms to transition retrenched workers and their dependents. This chapter chronologically continues with the narrative in the next phase of Japan's coal mining industry, which is post-closure gentrification of former coal mining areas. In this chapter, Lim utilizes his Hokkaido fieldwork as a case study of post-closure gentrification. The fieldwork was conducted with a generous grant from the Sumitomo Foundation for Japanese Studies.

The current stereotypical view of Hokkaido is a picture of scenic and picturesque wide spaces dotted with dairy farms. Many East Asian tourists yearn to visit Hokkaido to look at the wide open spaces, ski during winter (e.g. Katsurazawa-kokusetsu skiing area from December to March), sample high-quality dairy products from its farms, or/and stay in community inns (*ryokans*) to enjoy warm northern Japanese hospitality (*omakase*). The beautiful mountainous scenery is a major attraction for visitors and

T. W. Lim et al., *Coal Mining Communities and Gentrification in Japan*, https://doi.org/10.1007/978-981-13-7220-9_4

forms an important asset for Hokkaido tourism. The origins of these mountain ranges lie in ancient geological events. The clash of tectonic plates resulted in built-up pressures that elevated mountain ridges and ranges in Hokkaido, particularly in the central areas. Ancient forests that grew up in Hokkaido were eventually collapsed by tectonic forces and these powerful forces also exerted pressure and converted the dead trees into peat (through prolonged pressurization), from which coal emerged.[1]

Until Japan's domestic coal industry was mothballed, Hokkaido also had thriving coal mine towns. Yubari, Akabira, Utashinai, and Mikasa are all located in the coal-rich Sorachi region in Hokkaido. In the pre-modern period, Hokkaido's fishermen were well aware that Shiranuka and Kushiro were coal rich, but using coal resources to generate energy did not really take off until West-inspired modernity and modernization arrived in Japan. Eventually, in the early modern period, what catalysed the construction of a coal mining industry (especially for exports) in the region was the rise of Hakodate port for foreign trade, and the steamships that came from the West needed coal for replenishing supplies and facilities.[2] Modernity and its insatiable hunger for energy had arrived in Hokkaido. Foreign help was needed to develop the industry in Japan.

In 1871, then vice-Hokkaido Development Commissioner Kiyotaka Kuroda searched for inspirations from US economic development to apply in Hokkaido's own developmentalism; therefore, Horace Capron, who was the US Commissioner of Agriculture, was hired to be a consultant, project lead, and planner for the government. Based on Capron's ideas, the Hokkaido Development Commission designated and implemented the opening of the Horonai Coal Mine as a national priority.[3] This gave rise to the construction of the modern coal mining town of Mikasa. Within the Sorachi region, Mikasa's modern history began from the start of coal discovery in 1868 (year of Meiji Restoration, genesis of modern Japan)

[1] Taniguchi, Masaharu, "Mikasa Geopark Exploring the inner workings of our planet Earth" in the Hokkaido Magazine Kai [downloaded on 1 Jan 2017], available at http://kai-hokkaido.com/en/feature_vol32_mikasa/

[2] Kasuga, Yutaka, "Transfer and development of coal-mine technology in Hokkaido" dated 1982 in the Series: Japanese Experience of the UNU Human and Social Development Programme series; 48 [downloaded on 1 Jan 2017], available at https://d-arch.ide.go.jp/je_archive/pdf/workingpaper/je_unu48.pdf, p. 2.

[3] Hokkaido International Center, Japan International Cooperation Agency, Regional and Industrial Promotion Study in Hokkaido Study Report (Japan: Hokkaido International Center, Japan International Cooperation Agency, Hamanasu Foundation), 2015, p. 5.

and the authorities invested in the state-owned Horonai Coal Mine in 1879 (located in southeastern Mikasa) and, before long, Ichikishiri-mura Village, where contemporary Mikasa is located, was founded in 1882.[4]

Prison labour was used to develop the coal mines and the prisoners were compulsorily drafted from Mikasa's Sorachi-shujikan Prison founded in 1882 to work in the Horonai Coal Mine.[5] Contemporary local residents in that community are grateful for their pioneering work. The first state-owned coal mine here was established by the Horonai Coal Mine in 1879, on the cusp of Japan's modernization process. Japanese coal-powered steam engine railway history, a symbol of modernization, also started here as Benkeigo trains operated on the Horonai railway line (Japan's very first industrial railway system in 1882) to transport coal between Horonai and Temiya's port (Otaru City). Much of this industrial-age equipment remains in various conditions, attracting industrial heritage buffs and photographers to these sites. Industrial historical relics are by themselves a major tourist attraction, particularly amongst history buffs, railway enthusiasts (*densha-taku*), and industrial heritage fans.

In the course of the research for this edited volume, Lim joined *densha-takus* and industrial heritage buffs on such trips in July 2016 to curate and archive images of heritage artefacts found in this region. The predecessor of the Hokkaido railway line at Yubari today was the first industrial railway line in Japan. A heritage highlight for history buffs is a disused old passenger train with a seating capacity of 104 during summer and 96 during winter that is left standing at an abandoned railway track at the former Minamioyubari train station which is located next to a popular cycling track. The antiquated interior of the disused train at the former Minamioyubari train station with wooden floorboards attracts curious visitors, motorists, cyclists, photographers, and hikers into the cabin. Signs are posted inside the train that urge the public not to damage the two train carriages on display because the trains were workhorses that once ferried many residents to Yubari, Kagoshima, and Minamioyubari's coal mines. The idea is to respect the material artefact that once served commuters. The sign then urges visitors to relax slowly in the cabin and treat the cabins

[4] Mikasa City, "The City of Mikasa: a great place where you can enjoy!!" in the Mikasa City website [downloaded on 1 Jan 2017], available at http://www.city.mikasa.hokkaido.jp/hotnews/detail/00001591.html

[5] Mikasa City, "The City of Mikasa: a great place where you can enjoy!!" in the Mikasa City website [downloaded on 1 Jan 2017], available at http://www.city.mikasa.hokkaido.jp/hotnews/detail/00001591.html

Fig. 4.1 An old antiquated unused steam locomotive (SL) lies still on the railway track at the former Minamioyubari station. Built by Mitsubishi, it weighs 28.2 tons. (Photo taken by the author on 8 July 2016)

with great care as important industrial heritage. The disused railway track at the former Minamioyubari railway station also attracted many photography buffs in addition to history/heritage fans. The local community's love of heritage and sense of history is very strong in Yubari, as even old buses are preserved for local community and visitors alike to view and appreciate the recent Showa history (Fig. 4.1).

Other old equipment left at the site included the old signal, a PA (Public Announcement) loudspeaker equipment which was left at the site for heritage buffs, and a freight car that was used to carry coal and other minerals. The weight of the freight car is 14.9 tons and it can carry a coal load of 30 tons. The train car was manufactured by Mitsubishi. The hinges on the trough allow the doors to open outwards to unload coal after delivery. The public displays at Yubari were labelled with the official tags, marking the conserved "Heritage of Industrial Modernization" in modern Japan by the Keizai Sangyousho or Ministry of Economy, Trade and Industry (METI). Before starting discussions on the Hokkaido mines, the next literature review section examines some existing writings on this subject matter (Fig. 4.2).

Fig. 4.2 A 50-foot red colour head-house rigging standing next to the Yubari Coal Mine Museum has become a local landmark. (Photo taken by the author on 8 July 2016)

Literature Review[6] Japan's Hokkaido region is not alone in gentrifying former coal mines. Throughout East Asia, different mining regions are trying to cash in on their mothballed or active mine assets for tourism. Some former mines are rich in cultural heritage assets, others have man-made structures of interest to photography enthusiasts or history buffs, while some of them just feature laid-back rustic towns suitable for tourism. Local community leaders, and national and state governments have become aware of the potential of these sites to generate income for the local government of a province or state, but especially for the local community in that area. Therefore, vigorous efforts were made through East

[6] The minor section on non-Japanese gentrification of former mines is drawn from the author's blog publications: Lim, Tai Wei, "The State of Coal Industry in Northeast Asia" dated 10 August 2017 in IPP Review [downloaded on 10 August 2017], available at http://ippreview.net/index.php/Blog/single/id/527.html

Asian mining sites to conserve heritage assets, reproduce nostalgic lifestyles or working conditions, and/or gentrify the sites for retail and service industries. In Malaysia, the former tin mining town of Taiping is the most prominent example of a gentrified former mining area. It was transformed from a Wild Western–type frontier town that experienced horrific battles between rival secret societies formed by different groups of Chinese tin miners in the 1800s to a well-known regional tourist attraction. Besides the mothballed tin mining areas, Great Britain's administrators built heritage places of worships, parks, and European revival structures that are preserved for posterity and tourism.[7] The prominent sites include limestone structures, Lake Gardens' tropical primary forest rain trees, nature treks and picturesque locations, as well as Bukit Larut (Maxwell Hill)'s hiking trails, the Burmese pool waterfall, and wild jungles.[8] Still a laid-back bazaar town, two magnificent 130-year old wooden arcades are platforms for fruit retailers in daytime and for hawkers in evening time selling local fares like *kuay teow goreng* and *chee cheong fun*.[9]

Another less well-known Malaysian tin mine, Sungai Lembing, was prominent in the twentieth century for its wealth of subterranean tin reserves, with some of the biggest, lengthiest (300 kilometres), and deepest (700-metre-long tunnels) tin mines in the world being exploited by the Chinese employees of the British colonial authorities' Pahang Consolidated Company Limited (PCCL) in the early twentieth century.[10] With state funding, some touristy features in this gentrified mine included the Sungai Lembing Rainbow Waterfall, where a rainbow is visible in the

[7] Guardian writers, "Undiscovered south-east Asia: remote towns and secret beaches" dated 10 May 2017 in The Guardian [downloaded on 10 May 2017], available at https://www.theguardian.com/travel/2017/may/10/undiscovered-asia-thailand-vietnam-cambodia-laos-bali-beaches-islands

[8] Guardian writers, "Undiscovered south-east Asia: remote towns and secret beaches" dated 10 May 2017 in The Guardian [downloaded on 10 May 2017], available at https://www.theguardian.com/travel/2017/may/10/undiscovered-asia-thailand-vietnam-cambodia-laos-bali-beaches-islands

[9] Guardian writers, "Undiscovered south-east Asia: remote towns and secret beaches" dated 10 May 2017 in The Guardian [downloaded on 10 May 2017], available at https://www.theguardian.com/travel/2017/may/10/undiscovered-asia-thailand-vietnam-cambodia-laos-bali-beaches-islands

[10] Ong, Han Sean, "Sungai Lembing's evolution from a tin-mining hub to a tourist destination" dated 20 March 2014 in The Star Online [downloaded on 1 Jan 2017], available at http://www.thestar.com.my/news/community/2014/03/20/tunnels-from-a-glorious-past-sungai-lembings-evolution-from-a-tinmining-hub-to-a-tourist-destination/

water in the morning, a tin mining museum established in 2003, Panorama Hill's sunrise view, and Kuantan Municipal Council's refurbished mine shaft for tourist rides on a trolley car into the tunnel and walks.[11]

Indonesia is another location that is benefitting from a tourism boom in gentrified mining areas. A former gold mining area, Manado's scenic view with bamboo groves, rice fields and orchards, and a 50-metre-tall Christ's Blessing statue has inspired locals to take up tour guiding as a means to make a living, capitalizing on an Indonesia tourism boom that highlights outdoor adventure.[12] The trend has attracted both domestic tourists (cosmopolitan Jakarta residents who usually visit tourist-friendly areas) and mainland Chinese tourists (30,000 of them in 2016, projected to go up to 150,000 in 2017, boosted by increased flights) to Manado.[13] Other attractions in the Manado region that used to depend on gold mining and farming included Bunaken Island National Park reef in North Sulawesi for divers and snorkelers, and the region is treading carefully between sustainable tourism, unique local community Minahasan lifestyles, and commodified tourism.[14]

Another well-known but physically challenging area for tourists to traverse is the Ijen Crater in East Java. In the night, noxious gas combustion in the volcanic crater emits a "blue fire" that attracts hundreds of lined-up tourists as they navigate precariously downwards into the volcanic depression.[15] These selfie-taking tourists visit the site at the same time as the miners reporting to work at 2 am to escape the unbearable heat of the afternoon sun and

[11] Ong, Han Sean, "Sungai Lembing's evolution from a tin-mining hub to a tourist destination" dated 20 March 2014 in The Star Online [downloaded on 1 Jan 2017], available at http://www.thestar.com.my/news/community/2014/03/20/tunnels-from-a-glorious-past-sungai-lembings-evolution-from-a-tinmining-hub-to-a-tourist-destination/

[12] Raslan, Karim, "An Indonesian Gold Hunter Dreams of Better Days – in tourism" dated 2 March 2017 in the South China Morning Post (SCMP) website [downloaded on 2 March 2017], available at http://www.scmp.com/week-asia/business/article/2075441/indonesian-gold-hunter-dreams-better-days-tourism

[13] Raslan, Karim, "An Indonesian Gold Hunter Dreams of Better Days – in tourism" dated 2 March 2017 in the South China Morning Post (SCMP) website [downloaded on 2 March 2017], available at http://www.scmp.com/week-asia/business/article/2075441/indonesian-gold-hunter-dreams-better-days-tourism

[14] Raslan, Karim, "An Indonesian Gold Hunter Dreams of Better Days – in tourism" dated 2 March 2017 in the South China Morning Post (SCMP) website [downloaded on 2 March 2017], available at http://www.scmp.com/week-asia/business/article/2075441/indonesian-gold-hunter-dreams-better-days-tourism

[15] Board, Jack, "Asia's Toughest Jobs: The fire and brimstone miners of Ijen Crater" dated 19 August 2017 in the Channelnewsasia website [downloaded on 19 August 2017], available at http://www.channelnewsasia.com/news/asiapacific/asia-s-toughest-jobs-the-fire-and-brimstone-miners-of-ijen-9134662

also trade with the tourists amidst their numbers as additional revenue.[16] The miners wait for sulphur gas emitted by the volcanic crater to liquefy using a pipe system and then cool down to a solid form for the miners to extract and therefore, sometimes, tourists who show empathy for the miners purchase masks and filters (better protection than cloth over their mouths) for them using their own money.[17] Sunlight that penetrates through the low-altitude atmosphere generates warmth, painting the acidic lake in the volcanic crater with rainbow colours, stopping even the most hardened weary miners (in addition to tourists) to admire its beauty.[18]

Developing economies are not the only ones gentrifying older mines. Even within Japan, Hokkaido is not the only region gentrifying its former mines and railway tracks for tourism. In Gifu, for example, unused Kamioka railway tracks were adapted for a "mountain rail bike" tour system whereby tourists can ride mountain bicycles directly plying on the tracks to bring in tourist revenues and rejuvenate the greying local communities in the countryside.[19] The former freight-only railway tracks were used to transport mining products from Kamioka (contemporary Hida City, Gifu Prefecture), an eighth-century mine that was amongst the biggest zinc and lead ore mines in East Asia, and these minerals served Japan's economic development during the boom days before it was mothballed in 2006.[20] To support tourist arrivals and their activities within the former mining areas,

[16] Board, Jack, "Asia's Toughest Jobs: The fire and brimstone miners of Ijen Crater" dated 19 August 2017 in the Channelnewsasia website [downloaded on 19 August 2017], available at http://www.channelnewsasia.com/news/asiapacific/asia-s-toughest-jobs-the-fire-and-brimstone-miners-of-ijen-9134662

[17] Board, Jack, "Asia's Toughest Jobs: The fire and brimstone miners of Ijen Crater" dated 19 August 2017 in the Channelnewsasia website [downloaded on 19 August 2017], available at http://www.channelnewsasia.com/news/asiapacific/asia-s-toughest-jobs-the-fire-and-brimstone-miners-of-ijen-9134662

[18] Board, Jack, "Asia's Toughest Jobs: The fire and brimstone miners of Ijen Crater" dated 19 August 2017 in the Channelnewsasia website [downloaded on 19 August 2017], available at http://www.channelnewsasia.com/news/asiapacific/asia-s-toughest-jobs-the-fire-and-brimstone-miners-of-ijen-9134662

[19] Japan Times, "Tourists take part in a 'mountain rail bike' tour on the abandoned Kamioka Railway in Hida, Gifu" dated 23 May 2017 in the Japan Times website [downloaded on 23 May 2017], available at https://www.japantimes.co.jp/news/2017/05/23/national/efforts-afoot-turn-abandoned-rural-train-lines-tourist-attractions/#.WZo9oD4jGUk

[20] Japan Times, "Tourists take part in a 'mountain rail bike' tour on the abandoned Kamioka Railway in Hida, Gifu" dated 23 May 2017 in the Japan Times website [downloaded on 23 May 2017], available at https://www.japantimes.co.jp/news/2017/05/23/national/efforts-afoot-turn-abandoned-rural-train-lines-tourist-attractions/#.WZo9oD4jGUk

local/state government, private sector entities, and local communities need to work together closely to limit the numbers of tourist arrivals when necessary, protect natural assets around the mine to preserve its beauty, respect local communities' cultures and lifestyles, as well as educate all stakeholders on the idea of sustainable development. The idea is to benefit local communities first and then generate income for local area development while enhancing tourism branding for the country as a whole and impressing tourists with a well-planned educational tour experience. This complements the non-traditional nature of tourism in all its forms (including outdoor adventure tourism and cosmopolitan tourists keen on a local experience with an off-the-beaten track feel).

In terms of existing literature on the history of Hokkaido's coal mines, the most important writing on Hokkaido's late pre-modern to pre-war coal mining industry is Yutaka Kasuga's United Nations University (UNU) publication "Transfer and development of coal-mine technology in Hokkaido," where he detailed the history of Hokkaido's coal mines from the Tokugawa period onwards. A large part of this writing detailed the important chronological developments in surveying and eventually exploiting the land for its coal resources, including detailing early surveying efforts made by foreigners at the start of Japan's modernization. The writing then detailed the history of early infrastructure construction like the railway system, essential for the modernization of Hokkaido's coal mines. The use of prison labour, labour conditions, and manpower regulations were also discussed. The last section of Yutaka Kasuga's writing looked into pre-war Zaibatsu conglomerates and how their capital entered the Hokkaido coal industry during a phase of rationalization of operations when economy of scale mattered. This chapter's contribution to Yutaka Kasuga's writing is to examine the Hokkaido mines from the period of coal industrial decline starting from the 1960s to contemporary developments in former coal mines, including the processes of gentrification, small-scale production revival, tourism, environmental upgrading, and sustainable agriculture. This writing extends the pre-modern context in Yutaka Kasuga's writing to the present day.

Existing literatures on this subject at the point of writing this chapter are mostly policy papers instead of an edited volume. For comparative references, general policy reports on other coal mine closures in the world that include writings such as *Research on Mine Closure Policy*[21] are utilized as

[21] Cochilco, Chilean Copper Commission, "Research on Mine Closure Policy No. 44" dated January 2002 in the International Institute for Environment and Development (IIED) website [downloaded on 1 Jan 2017], available at http://pubs.iied.org/pdfs/G00541.pdf

secondary sources in this project for conceptual ideas to formulate a coherent and consolidated study. Articles and academic studies that date back to the 1960s were also used in order to understand the subject matter from a temporal standpoint perspective drawn from that historical period. Some literatures on coal mine closures were also examined. For example, Benjamin Martin's article detailed the social impact of an early mine closure from the perspective of the early 1960s.[22] Another classic article "Some Problems of Unemployed Laborers in Coal Mining Industry" by Naoki Kusuhara[23] details the first post-war attempt to address post-closure issues such as re-employment of retrenched workers. It is important to note that these are short articles rather than a consolidated coherent edited volume structured around a central theme. The most important monographic work in the English language is: Culter, Suzanne, *Managing Decline: Japan's Coal Industry Restructuring and Community Response* (Hawaii: University of Hawaii Press), 1999. It is mainly written from a sociological/anthropological point of view and with information dated to the 1999. Many gentrification projects in closed mines thereafter are not discussed in the volume.

In Lim Tai Wei's previous publication (*Energy Transitions in Japan and China: Mine Closures, Rail Developments, and Energy Narratives*), Lim studied the research question of why coal mines declined in importance in the post-war years. In that publication, Lim focused mainly on mines found in southern Japan, particularly in Kyushu, like Omuta, Chikuhō, and other mines that have since closed when Japan transitioned to the use of oil starting from the 1960s. In the course of his previous research, he came across information related to the gentrification of these former mines and discovered how they have been transformed into environmentally friendly towns with smaller populations. The former mine sites have also been converted into playgrounds, skating parks, museums, and other educational facilities. Up till that point of time, literatures on the gentrification of former coal mines were limited, with Suzanne Cutler's publication *Managing Decline: Japan's Coal Industry Restructuring and Community Response* as the major seminal work in this area. This writing is a continuation of Lim's previous work on former coal mines in Japan. There are

[22] Martin, Benjamin, "Japanese Mining Labor: The Miike Strike" dated Feb 1961 in Far Eastern Survey Vol. 30 No. 2 (American Council, Institute of Pacific Relations), 1959, pp. 26–30.

[23] Kusuhara, Naoki, "Some Problems of Unemployed Laborers in Coal Mining Industry" in The science reports of the Tohoku University 7th series, Geography Vol. 17 No. 1 (Japan: Tohoku University), 1968, pp. 105–116

some distinguishing features of this volume from previous works. In Lim Tai Wei's previous volume, he mainly worked on coal mines in southern Japan, Kyushu. In this writing, he carried out fieldwork with his Japanese colleagues in sites of northern coal mines in Hokkaido like Kushiro, Yubari, and Bibai.

Objectives, Framework, and Methodology The writing's central focus complementing the edited volume's objectives is analysing the agricultural cooperatives, local environmental movements/initiatives, archived memories, and community-based initiatives in former coal mining towns and local economies in Japan. In terms of methodology, this writing utilizes an eclectic approach that includes textual interpretation of secondary sources, observation studies during field visits, and visits to archives related to the subject matter in Japan. More details are appended in the section below. This process took place over two years, from 2016 to 2017, and thereafter taking another year for writing and analysis of gathered materials and curated materials. It follows up on the author's previous work (Lim, Tai Wei, *Energy Transitions in Japan and China Mine Closures, Rail Developments, and Energy Narratives* (UK: Palgrave Macmillan), 2017) which mainly looks at Japanese shuttered coal mines in the south of Japan (Kyushu), with comparative references to China.

This writing utilized observation studies to go on-site to observe how gentrified coal mining spaces are used specific to the location of Hokkaido, a coal mining region of historical importance. In the process, the author spoke to individuals who were affected by coal mine closure or residents of these towns who were displaced and/or economically affected by the decline of this industry. In the museums, the author's observations also recorded the artefacts on display on digital images to curate how memories of the past are archived and how they continue to serve as educational tools for community residents. The author visited some of these mines to have a personal experience of life as a coal miner in order to adopt a standpoint view of entering and working in the coal mines.

The interpretive component relied on collection of textual documents (both primary and secondary) to reconstruct a clearer picture of the former coal mining towns before and after closure. These textual and visual documents were useful to corroborate information received from on-site observations. They can provide some empirical and quantitative data useful for the volume. Blueprints and town plans also provide a clearer spatial layout of the former coal mining facilities in the places that were visited by the author. The author selectively visited and conversed with users of gentrified green spaces, farmers in former coal mining towns, and local museum curators to understand their opinions of post-closure changes. The conversations are mostly in the form of open unstructured questions for local community individuals to articulate their views. For the author's observation studies, he divided the research work into two sectors.

In the summer of 2016, he visited West Hokkaido mines such as Yubari, Mikasa, and, very briefly, Bibai. Some of the sites the author visited in West Hokkaido in 2016 included: Bibai, Asahikawa, or Yubari Coal Mining Museums, cycling track and green rail corridor built on abandoned railway tracks in Kushiro, a homestay inn near an agricultural area in a former coal mining town in Hokkaido, Yubari film festival site and farming area, local area book shops, archives, and libraries in former coal mining towns in Hokkaido like Bibai, Asahikawa, or/and Yubari. The author joined the same Japanese team to visit East Hokkaido coal mines, including Kushiro, Urahoro, and Onbetsu from 14 July 2017 to 16 July 2017.

Modern and Contemporary History The late modern period saw a rapid rationalization of the coal mining industry in Japan. Toshitaka Nagahiro's important essay indicated that, in 1921, Japanese coal producers in Hokkaido and other coal mining regions formed the Sekitan Kôgyô Rengôkai (*Sotan Seigen* for short; Federation of Coal Mine Owners or FCMO) to work on stabilizing coal prices through managing demand and supply of coal resource by permitting each mine to produce fixed amounts of coal output.[24] Rationalization strengthened considerably in the pre-war years. In the 1930s, coal production quota for individual oil fields was fixed when each coal mine sent representative councillors (usually from large companies, Hokkaido typically had nine reps) to FCMO to negotiate and discuss quota compromises amongst themselves.[25] Rationalization

[24] Toshitaka, Nagahiro, "Japan's Coal Cartels in the Interwar Period" dated 17 March 2014 in the World Business History Conference (WBHC) website [downloaded on 1 Jan 2017], available at http://www.worldbhc.org/files/full%20program/A6_B6_JapanscoalCartelsintheinterwarperiod_ToshitakaNagahiro_.pdf

[25] Toshitaka, Nagahiro, "Japan's Coal Cartels in the Interwar Period" dated 17 March 2014 in the World Business History Conference (WBHC) website [downloaded on 1 Jan 2017], available at http://www.worldbhc.org/files/full%20program/A6_B6_JapanscoalCartelsintheinterwarperiod_ToshitakaNagahiro_.pdf

was necessary because coal mining regions were growing rapidly in terms of output, population, and urban centres. For example, Yubari was established in 1888 when a coal seam was discovered.[26] Yubari prospered economically alongside the Ishikari Coal Mine, which was exploited starting from 1891, and its population grew rapidly and, before long, it was designated a city in 1943.[27]

After WWII, Yubari's post-war development took off from the 1950s when 24 coal mines were exploited by Hokkaido Colliery & Steamship (Hokutan) and Mitsubishi, with trickle-down effects to other coal-related sectors.[28] In the 1950s, Yubari had 24 coal mine sites organized into 15–20 coal mine districts, with 5000–15,000 residents each, and was the biggest coal mine city in Japan, but an ageing population has seen a steady migration from North (mainly mining areas) to the agriculturally developed and touristy South Yubari.[29] As coal became increasingly important as a source of energy in the modern period, the coal mines are extracted for greater volumes of coal output, powering industrial applications and eventually war machines. In 1940, Japanese domestic coal mines achieved their peak production with an output of 56.31 million tons of coal,[30] and after 1945, it was identified alongside shipbuilding, steelmaking, and automobile production as a priority industry in immediate post-war Japan.

[26] Seaton, Philip, "Depopulation and Financial Collapse in Yūbari: Market Forces, Administrative Folly, or a Warning to Others?" dated Winter 2010 in the Social Science Japan Journal Vol. 13 No. 2 [downloaded on 1 Jan 2017] (UK: Oxford University Press), available at https://www.jstor.org/stable/pdf/40961265.pdf?refreqid=excelsior:db749056af5c857ae4ec24e37b1de1cf, pp. 229–230

[27] Hokkaido International Center, Japan International Cooperation Agency, Regional and Industrial Promotion Study in Hokkaido Study Report (Japan: Hokkaido International Center, Japan International Cooperation Agency, Hamanasu Foundation), 2015.

[28] Hokkaido International Center, Japan International Cooperation Agency, Regional and Industrial Promotion Study in Hokkaido Study Report (Japan: Hokkaido International Center, Japan International Cooperation Agency, Hamanasu Foundation), 2015.

[29] Yanagida, Ryozo, "Yubari: Revitalization of coal mining city" dated September 2006 in the Meiji Gakuin University website [downloaded on 1 Jan 2017], available at http://www.meijigakuin.ac.jp/~hattori/shrinkingcity/page/JAPAN/yuubari-slimmed-103208k.pdf

[30] Yoshimura, Yoshito and the Coal Division Agency of Natural Resources and Energy (Japan), "Closure of Japan's Last Large-Scale Coal Mine" dated 15 March 2002 in the Institute for International Studies and Training (IIST) World Forum website [downloaded on 1 Jan 2017], available at http://www.iist.or.jp/wf/magazine/0067/0067_E.html

Japan's transition to oil as an energy resource, desire for reduction of pollution from coal burning, and rising manpower costs contributed to the decline of its domestic coal mining industry. Rising Japanese yen also made domestic coal more expensive than foreign-imported coal by the 1970s, and by the time Kushiro coal mines were closed, Japanese coal was three times more expensive than foreign imports.[31] By 1980, there was a significant difference between the cost of importing coal for power plants (US $52 a ton, inclusive of freight costs) and domestic coal (US $77), with price increase slowed down because of productivity gains.[32] By 2008, the cost of producing a ton of Japanese domestic coal was at least US $100.[33] Moreover, the advent of alternative energy sources, hybrid vehicles, fuel cells, and renewable energy sources that are cleaner and more fuel efficient further affected the viability of having a domestic coal industry in Japan. Experimental energy sources were also attempted. When the coal mines shut down in Bibai, approximately 80 firms or more took part in an electricity generation project using coal bed methane (CBM), carbon dioxide, and a snow-cooling system based on natural energy (cryogenic energy of snow) (Fig. 4.3).[34]

When mines in Hokkaido started closing (a delayed outcome that had already taken place in mines in other parts of Japan), an expert in the study of rehabilitation of coal mine workers, Naoko Shimazaki, highlighted how the Hokkaido authorities worked with other stakeholders to transfer retrenched coal workers to other mines (a process known as *tanko-fukki* or "returning to mining work" when translated into the English language), especially to the state-invested expanding "build mines," which were experiencing human resource shortages, for example, the Mitsubishi Minami Oyubari Mine "build mine" founded in 1966 that absorbed 1540

[31] Yoshimura, Yoshito and the Coal Division Agency of Natural Resources and Energy (Japan), "Closure of Japan's Last Large-Scale Coal Mine" dated 15 March 2002 in the Institute for International Studies and Training (IIST) World Forum website [downloaded on 1 Jan 2017], available at http://www.iist.or.jp/wf/magazine/0067/0067_E.html

[32] Lohr, Steve, "Japan's Big Energy Setback" dated 27 Oct 1981 in The New York Times website [downloaded on 1 Jan 2017], available at http://www.nytimes.com/1981/10/27/business/japan-s-big-energy-setback.html

[33] Fackler, Martin, "As Oil Prices Rise, Nations Revive Coal Mining" dated 22 May 2008 in The New York Times [downloaded on 1 Jan 2017], available at http://www.nytimes.com/2008/05/22/business/worldbusiness/22mines.html?mcubz=0

[34] Hokkaido International Center, Japan International Cooperation Agency, Regional and Industrial Promotion Study in Hokkaido Study Report (Japan: Hokkaido International Center, Japan International Cooperation Agency, Hamanasu Foundation), 2015

Fig. 4.3 The Yubari mine entrances were not grilled up and left open but were overgrown with vines, blending into the natural background. This was again a typical example of nature reclaiming its place over manmade structures. Others are closed to the public, with the rusted grills of the mothballed coal mine in Yubari preventing entry. (Photo taken by author on 8 July 2016 at Yubari)

(at the height of its record employment) redundant coal miners from other mines that were closed down.[35] Studying developments in Hokkaido coal mines during this period is important because the core centre of Japanese coal mining activities had moved from Kyushu in southern Japan to Hokkaido in the North in the 1960s, a transition that lasted until the closure of the last major mine in Japan (Fig. 4.4).[36]

[35] Shimazaki, Naoko, "Support for Workers Displaced in the Decline of the Japanese Coal Industry: Formal and Informal Support" available at http://www.jil.go.jp/english/JLR/documents/2015/JLR46_shimazaki.pdf, p. 11.

[36] Shimazaki, Naoko, "Support for Workers Displaced in the Decline of the Japanese Coal Industry: Formal and Informal Support" available at http://www.jil.go.jp/english/JLR/documents/2015/JLR46_shimazaki.pdf, p. 18.

Fig. 4.4 The eastern area of Yubari City (Oyubari district) has given way to a hydroelectric dam. (Yanagida, Ryozo, "Yubari: Revitalization of coal mining city" dated September 2006 in the Meiji Gakuin University website [downloaded on 1 Jan 2017], available at http://www.meijigakuin.ac.jp/~hattori/shrinkingcity/page/JAPAN/yuubari-slimmed-103208k.pdf.) This is a photograph of a hydro-electric power dam near Yubari, a good example of renewable clean sources of energy that has superseded the need for more expensive domestically produced coal in Japan. Another source of renewable energy is solar power. When the author landed at Chitose airport, he spotted advertisements put up by Japanese solar panel companies operating in Hokkaido. Memories of historical structures in the dam are well conserved through heritage signage, which has completely covered the former Sangengyou railway bridge that was located here. It is now found within the dammed-up water. (Photo taken on 8 July 2016)

Small-Scale Production Despite the closure of coal mines in Japan, at the back of the minds of some coal advocates lies the persistent idea that Japanese coal deposits are still substantial and there is a deep-seated desire to reduce overreliance on imported sources of energy, particularly after the 1973 and 1979 oil crises. Sorachi (where Bibai is located) has about six billion tons of coal, sufficient to meet Japan's current needs for three decades.[37] In some

[37] Fackler, Martin, "As Oil Prices Rise, Nations Revive Coal Mining" dated 22 May 2008 in The New York Times [downloaded on 1 Jan 2017], available at http://www.nytimes.com/2008/05/22/business/worldbusiness/22mines.html?mcubz=0

areas, small-scale coal mining activities continued. The town of Bibai first got into international prominence recently when it was featured in *The New York Times* as an example of the gentrification process and industrial revival in Japan in 2008. At the height of oil prices then, Bibai made a decision to go into coal production once again after Japan transited from coal to oil energy. The Taiheiyo Coal Mine in Kushiro was Japan's very last large coal mine that closed in January 2002, but inhabitants of Kushiro founded the Kushiro Coal Mine to continue small coal extraction operations starting from April 2002. According to the Jiji News agency, Kushiro miners continued to descend into the depths of the Kushiro Coal Mine (150 metres below sea level and Japan's only underground coal mine still being exploited) in a cable car, and because of such efforts in 2014, approximately 5000 tons of coal is extracted daily for use in power utility plants.[38]

Interestingly, whenever global oil prices went up or when energy supplies were affected because of wars and other events, small-scale coal production boomed in Hokkaido. On Thursday, 22 May 2008, when oil prices soared more than US $135 a barrel, domestic production at Bibai's two remaining strip mines reached their highest outputs in nearly 40 years when production levels went over 150,000 tons in 2008 (up from 34,961 tons in 2005), which is a record figure since 1973 when the last major underground Bibai mine closed.[39] In the fateful year of 2008, Hokkaido Electric Power acquired twice the amount of Japanese coal (up to 110,000 tons) compared to the previous year to cope with the strong demand; for example, Mitsubishi Materials cement producer purchased domestic Japanese coal in 2008 (something it had not done in 18 years prior to 2008), and to cope with such demand, the Hokuryo mine was estimated to have extracted 120,000 tons of coal while expanding production to a second mine in the same period.[40]

[38] Jiji Press Photo, "Japan Coal Mine" dated 8 August 2014 in the Jiji Press Photo website [downloaded on 1 Jan 2017], available at http://jijiphoto.photoshelter.com/image/I0000WV746N7o9aU

[39] Fackler, Martin, "As Oil Prices Rise, Nations Revive Coal Mining" dated 22 May 2008 in The New York Times [downloaded on 1 Jan 2017], available at http://www.nytimes.com/2008/05/22/business/worldbusiness/22mines.html?mcubz=0

[40] Fackler, Martin, "As Oil Prices Rise, Nations Revive Coal Mining" dated 22 May 2008 in The New York Times [downloaded on 1 Jan 2017], available at http://www.nytimes.com/2008/05/22/business/worldbusiness/22mines.html?mcubz=0

After the occurrence of the 2011 Great East Japan Earthquake (known popularly as "311"), Hokkaido's Ishikari Coal Mine (which also extends into Bibai), which was at one point the largest coal mine nationwide, bounced back into production again and its coal output in the post-311 years made up for the stoppage of Hokkaido's two nuclear power plants that previously supplied 40% of the prefecture's electricity power needs before the triple disaster (earthquake, tsunami, and nuclear meltdown in 311).[41] The coal mines at Bibai were also hard at work supplying Hokkaido Electric coal drawn from maximum production at its Nishimukizawa strip mine.[42] While there are growth spurts in coal production when global energy prices are high, the caveat is that these are likely to be small-scale productions that are revived only when oil prices are high and with the occurrence of other external shocks like 311 Great East Japan Earthquake. The long-term decline of the Japanese domestic coal industry is in many ways irreversible. Manpower in this declining industry is likely to be a limiting factor as well. In terms of expertise, according to *The New York Times*' Martin Fackler, there were hardly any coal specialist geologists left in Japan by 2008.[43] Moreover, coal mining is considered a three "K" job (*kitsui*, *kitanai*, and *kiken* or demanding, dirty, and dangerous) that no local Japanese is keen on taking up.

The dangers of working in a coal mine were dramatically demonstrated in 1981. In 1981, the Yubari Coal Mine managed by the Kokutan Yubari Mining Company had an output of one-million ton annually out of Japan's then total domestic output of approximately 18 million tons or one-fifth of the overall coal volume Japan uses annually.[44] An unfortunate accident occurred that year. The Yubari Coal Mine mishap in 1981 where

[41] Japan Times, "Nuke crisis a boon for struggling coal mines" dated 31 March 2012 in the Japan Times website [downloaded on 1 Jan 2017], available at http://www.japantimes.co.jp/news/2012/03/31/national/nuke-crisis-a-boon-for-struggling-coal-mines/#.WUXDzOuGOpo

[42] Japan Times, "Nuke crisis a boon for struggling coal mines" dated 31 March 2012 in the Japan Times website [downloaded on 1 Jan 2017], available at http://www.japantimes.co.jp/news/2012/03/31/national/nuke-crisis-a-boon-for-struggling-coal-mines/#.WUXDzOuGOpo

[43] Fackler, Martin, "As Oil Prices Rise, Nations Revive Coal Mining" dated 22 May 2008 in The New York Times [downloaded on 1 Jan 2017], available at http://www.nytimes.com/2008/05/22/business/worldbusiness/22mines.html?mcubz=0

[44] Lohr, Steve, "Japan's Big Energy Setback" dated 27 Oct 1981 in The New York Times website [downloaded on 1 Jan 2017], available at http://www.nytimes.com/1981/10/27/business/japan-s-big-energy-setback.html

93 miners perished was the deadliest since 1963 when 237 perished in another accident.[45] After the outbreak of the accident in 1981, the company that ran the Yubari Coal Mine inserted 12,751 tons of water into a mine to extinguish a fire that resulted from a gas explosion on 16 October 1981, which killed a total of 93 coal miners (54 of them) and emergency personnel (49 individuals) who were trapped inside the coal mine saturated with toxic gas.[46] To recover bodies and locate the unaccounted individuals, emergency personnel had to first repair the fans and damaged areas to facilitate new airflow into the tunnels again.[47] Disasters like these were often objects of sensational and high-profile media coverage that cemented negative public opinions on operating such dangerous facilities. After the 1981 accident, Hokutan Yubari Tanko, one of the biggest components of and the most prominent entity under the Hokutan conglomerate, became insolvent.[48]

West Hokkaido: Post-Closure Environmental Considerations The bulk of the author's fieldwork in West Hokkaido focused on Yubari and Mikasa. Unused coal mines were mothballed and made defunct. To green-ify these former coal mines, a number of stakeholders are coming up with ways to make their environments conducive for tourism and local community livelihoods, ensuring a clean environment and sustainable development. One environmental initiative is to revegetate the coal mines in Hokkaido, especially the areas around the coal heaps where the coal slags were deposited when the mines were abandoned or mothballed. Without revegetation, contamination by metals found in the slags may damage the environment. Sometimes the bald parts of the coal heaps reveal black coal sections that

[45] Lohr, Steve, "Japan's Big Energy Setback" dated 27 Oct 1981 in The New York Times website [downloaded on 1 Jan 2017], available at http://www.nytimes.com/1981/10/27/business/japan-s-big-energy-setback.html

[46] UPI, "Operators of a Hokkaido coal mine today completed pumping" dated 26 Oct 1981 in the UPI website [downloaded on 1 Jan 2017], available at http://www.upi.com/Archives/1981/10/26/Operators-of-a-Hokkaido-coal-mine-today-completed-pumping/1687372920400/

[47] UPI, "Operators of a Hokkaido coal mine today completed pumping" dated 26 Oct 1981 in the UPI website [downloaded on 1 Jan 2017], available at http://www.upi.com/Archives/1981/10/26/Operators-of-a-Hokkaido-coal-mine-today-completed-pumping/1687372920400/

[48] Hokkaido International Center, Japan International Cooperation Agency, Regional and Industrial Promotion Study in Hokkaido Study Report (Japan: Hokkaido International Center, Japan International Cooperation Agency, Hamanasu Foundation), 2015

are unrestrained by vegetation. These bald parts can leak metallic elements into the natural environment when there is rain or other weather elements. During his fieldwork on 8 July 2016, the author spotted partial revelation of bald parts (blackish in colour) of the coal heap in Yubari while revegetation carried out by the authorities is transforming these coal heaps into environmentally green spaces. The green areas also extend to water bodies in the region. During the fieldwork on 8 July 2016, the author passed by rivers and streams near Yubari. Hokkaido has become an important source of potable water and also a site for river sports like rafting for nature enthusiasts.

In Yubari, a primary school that had ceased its operations was converted into a "nature academy" for urban children to try their hands at paddling down Yubari rivers and snare stag beetles for rearing in plastic boxes.[49] Like Yubari, Bibai similarly offers green spaces for visitors' activities, such as the Miyajima marsh wetland covered under the Ramsar Convention Treaty where tourists go to view the migratory greater white-fronted goose.[50] In terms of constructed natural space, Bibai hosts two wards of Toumeicho, which has become well known as a Sakura flower (cherry blossom)–viewing spot within the Sorachi region, with 2000 cherry blossom plants, 5000 azalea flowers, a proximate open-air cave-type hot spring *onsen*, and a golf course.[51]

Tourism Along with coal heaps, industrial infrastructures are also left behind in former coal mines, becoming another source of tourism assets for visitors. Previously perceived as rusting eye sores, after they were repainted and refurbished, some of these structures proved to be very popular with visitors and tourists. Some of these infrastructures are massive and can pique the interest of visitors who are interested in learning about modernization, industrialization, and mining activities through material artefacts.

[49] The Guardian, "Yubari, Japan: a city learns how to die" dated 15 August 2014 in The Guardian [downloaded on 1 Jan 2017], available at https://www.theguardian.com/cities/2014/aug/15/yubari-japan-city-learns-die-lost-population-detroit

[50] Hokkaido Bibai City Portal site, "Arte Piazza Bibai" dated 2008 in the PiPa Hokkaido Bibai City Portal website [downloaded on 1 Jan 2017], available at https://www.pipaoi.jp/fl/en/artepiazzabibai/

[51] Japan Travel and Tourism Association, "Bibai City Tomei Park" dated 2014 in The Must-see Japan website [downloaded on 1 Jan 2017], available at http://www.nihon-kankou.or.jp.e.wp.hp.transer.com/sakura/searchDetail.jsp?ken__=hokkaido&shi__=01215&id__=S0127

Others structures are officially labelled by the authorities as "Heritage of Industrial Modernization," and so it is a thrill for history buffs to hike or cycle to these official sites. For example, the Honbetsu Coal Mine also had a large tower structure (largest in Asia when it was constructed in 1960) that used to lower coal miners to the bottom of the pit.[52] Yubari is also a heritage site (in terms of both memories and museum artefacts) of early industrial technologies installed in Japan; for example, in 1902, an electric endless rope manufactured by the Temiya Manufacturing Plant was laid down from the Tenryu pit to Yubari's coal screening site, while some of the first electric fans in mines were constructed in Yubari in 1898.[53] Some of the earliest coal mine engineering schools in Japan were located in Yubari; for example, in 1920 Hokutan Yubari Engineering School offered three-year courses on coal extraction topics for an intake of 30 new students annually, eventually graduating 409 students over a period of 13 years.[54]

However, in the areas of tourist infrastructure, Yubari is proceeding carefully with its tourism initiatives, as it became financially insolvent in 2007 after constructing an entertainment theme park and a Coal History Village that proved unpopular with visitors and turned out to be white elephants.[55] Japan's Internal Affairs and Communications Ministry is overseeing Yubari's recovery from bankruptcy through an 18-year repayment schedule that runs until 2024, and the plan is dependent on greater contribution to state coffers from residential/property taxes and sewerage fees to shave off the city's 35.3 billion yen debt.[56] Amenities like the

[52] Taniguchi, Masaharu, "Mikasa Geopark Exploring the inner workings of our planet Earth" in the Hokkaido Magazine Kai [downloaded on 1 Jan 2017], available at http://kai-hokkaido.com/en/feature_vol32_mikasa/

[53] Kasuga, Yutaka, "Transfer and development of coal-mine technology in Hokkaido" dated 1982 in the Series: Japanese Experience of the UNU Human and Social Development Programme series; 48 [downloaded on 1 Jan 2017], available at https://d-arch.ide.go.jp/je_archive/pdf/workingpaper/je_unu48.pdf, p. 36.

[54] Kasuga, Yutaka, "Transfer and development of coal-mine technology in Hokkaido" dated 1982 in the Series: Japanese Experience of the UNU Human and Social Development Programme series; 48 [downloaded on 1 Jan 2017], available at https://d-arch.ide.go.jp/je_archive/pdf/workingpaper/je_unu48.pdf, p. 59.

[55] Fackler, Martin, "As Oil Prices Rise, Nations Revive Coal Mining" dated 22 May 2008 in The New York Times [downloaded on 1 Jan 2017], available at http://www.nytimes.com/2008/05/22/business/worldbusiness/22mines.html?mcubz=0

[56] Takahara, Kanako, "Yubari tries downsizing to solve fiscal crisis" dated 1 March 2007 in the Japan Times [downloaded on 1 Jan 2017], available at http://www.japantimes.co.jp/news/2007/03/01/national/yubari-tries-downsizing-to-solve-fiscal-crisis/#.WUoHu-uGOUk

library, the museum, and seven primary and four junior highs have been shut down, with 152 local officials resigned by March 2007 (including 54 of 57 senior officials).[57] Upon implementing these measures, civil society and non-profit organizations (NPOs) immediately kicked into action as they started to source for their own finances to organize the city's traditional events and festivals like the world-famous film festival (estimated 100 million yen)[58] that originated in 1989.[59]

The former Yubari Coal Mining Museum has shut down, but its surrounding heritage structures remain. In the past, the museum gave members of the public access down steps to descend into the depths of the coal mine. Here visitors could see steel structures supporting tunnels along with wooden shafts. Certain sections of the tunnel were reinforced with bricks. There were life-sized models of coal miners tending to equipment inside the coal mine. The coal mine facility had signs that labelled various equipment and explained each step of the coal production process, including pumps, mechanized coal shaft support, exhaust fans, and so on. It was a realistic mock-up of activities within an actual underground coal mine shaft. A Shinto shrine sits at the top of the coal mine entrance and the flag of Kushiro Coal Mine (KCM) flies over the ground of the museum. The shrine traditionally blessed coal miners before they entered the mines. We received special permission to enter the mothballed mines for experiential learning (Fig. 4.5).

Despite the closure of the coal mining museum with access to an actual coal mine shaft and the coal historical village, another red-bricked coal museum located nearby remains operational in Yubari. It features exhibits that include a simulator that reconstructed the atmosphere and feeling of being 1000 metres underground, just like the coal miners felt. Equipment like the German-made S3-Resshaken-Plough previously used in the Japanese coal mines were also on display. Botanical specimens like the

[57] Takahara, Kanako, "Yubari tries downsizing to solve fiscal crisis" dated 1 March 2007 in the Japan Times [downloaded on 1 Jan 2017], available at http://www.japantimes.co.jp/news/2007/03/01/national/yubari-tries-downsizing-to-solve-fiscal-crisis/#.WUoHu-uGOUk

[58] Takahara, Kanako, "Yubari tries downsizing to solve fiscal crisis" dated 1 March 2007 in the Japan Times [downloaded on 1 Jan 2017], available at http://www.japantimes.co.jp/news/2007/03/01/national/yubari-tries-downsizing-to-solve-fiscal-crisis/#.WUoHu-uGOUk

[59] Yanagida, Ryozo, "Yubari: Revitalization of coal mining city" dated September 2006 in the Meiji Gakuin University website [downloaded on 1 Jan 2017], available at http://www.meijigakuin.ac.jp/~hattori/shrinkingcity/page/JAPAN/yuubari-slimmed-103208k.pdf

Fig. 4.5 Near the closed museum is a large defunct coal mining tower painted red with a sign indicating the former coal mine historical village. (Photo taken by the author on 8 July 2016)

dawn redwood whose fossilized version created coal were also planted around the museum. Other museum exhibits included human figurines depicting coal miners in their uniform and with equipment, and an air compressor acquired from the US-based Sullivan Brothers Co. Ltd. used in pumps, locomotives, and picks.

Around the museum, there were also textual explanation signage of how the prehistoric age created coal in Japan and its replacement by oil, oil paintings related to the coal mine, plates that explained how coal was used in Japan to generate electricity and the possibility of its liquefaction, chemical qualities and gasification, and so on. Photos on display were also an important visual snapshot of life in the coal mines. These photos portrayed scenes of ***matsuri*** festivals (like Yama no Bonodori), the misty appearance of the Yubari mine in the morning, coal miners reaching up to 700 metres in depths for extracting coal, relaxed faces of coal miners taking a break, miners drilling the coal bed, miners' bodies and faces darkened with soot taking a bath, and so on. The most emotionally charged photos were those of a major accident that happened on 16 October 1981.

Museum-ification Some of these relics and building structures are professionally curated for visitors to the city. The Mikasa City Museum hosts Japan's largest collection of ammonite fossils (600 Cretaceous-period items belonging to 80 species and all found in the area), which are easily found in Hokkaido, and also owns some 100-million-year-old fossils of crustaceans, fish, botanical specimens, and a fossilized skull of a Yezo-mikasa-ryu (Mosasaurids; Cretaceous carnivorous marine reptiles), all of which are important artefacts that qualify the museum as a National Natural Monument.[60] All these lifeforms in the local area are accompanied by geological displays, including coal beds and mothballed coal mines.[61] Because of its collection, the Mikasa City Museum is currently nicknamed the "town of fossils" (ammonite artefacts and all other fossils added together come up to 1000 items), symbolized by the giant Bivalvia Inoceramidae fossil, which has been designated a national monument, and the Yezosaurus fossil uncovered in Mikasa in 1976.[62] Visitors are allowed to feel and touch majority of the fossils and imagine life in prehistoric time while participating in activities like "Nature Observation Courses" to visualize fossils in open spaces, in addition to viewing the local historical displays depicting Mikasa as the inaugural coal town in Hokkaido.[63]

[60] Go Musee Japan, "Mikasa City Museum" in the Go Musee website [downloaded on 1 Jan 2017], available at http://go-musee.com/archives/general/221/

[61] Go Musee Japan, "Mikasa City Museum" in the Go Musee website [downloaded on 1 Jan 2017], available at http://go-musee.com/archives/general/221/

[62] Hokkaido Tourism Organization (Hokkaido Department of Policy Planning and Coordination, Bureau of Transportation Policy, Transportation Policy Planning Division), "The area surrounding Sapporo, Otaru and Noboribetsu (Central Hokkaido)" in the Hokkaido Tourism Organization website [downloaded on 1 Jan 2017], available at http://shinkansen-access.visit-hokkaido.jp/en/area/area04/spot0408.html

[63] Hokkaido Tourism Organization (Hokkaido Department of Policy Planning and Coordination, Bureau of Transportation Policy, Transportation Policy Planning Division), "The area surrounding Sapporo, Otaru and Noboribetsu (Central Hokkaido)" in the Hokkaido Tourism Organization website [downloaded on 1 Jan 2017], available at http://shinkansen-access.visit-hokkaido.jp/en/area/area04/spot0408.html

Mikasa is not the only former coal mining city that has museum-ified its coal industrial heritage. In Kushiro, the Kushiro City Museum is an attraction for visitors with exhibits on the Kushiro swamps, seaweeds and sandy beaches, geological formations, Ainu minority cultural artefacts and prehistoric lifeforms (Jomon Neolithic culture), and the study of the Japanese crane (a migratory bird that stops over at Yubari).[64] Besides visitors travelling by themselves using maps in their guidebooks, there are professionally planned Sorachi region coal mine heritage tours ("The Coal Mine Experience Tour") organized by the NPO Coal Mining Remembrance Promotion Agency and Sorachi Coal Mining Remembrance Management Center that focus on the history of the Yama mountain in the Sorachi coal mines, which were important industrialization sites and are officially recognized as Hokkaido heritage sites in 2001 (e.g. Mikasa's Former Sumitomo Ponbetsu Coal Mine).[65] Some of these tours and activities are set up by grassroots initiatives such as the Akabira Community Guide Club Tantan and the NPO Coal Mining Remembrance Promotion Agency, established in 2007 to cater to increasing numbers of interested visitors and tourists (Fig. 4.6).[66]

Besides older structures, an unofficial shrine was built at the location to appease the souls of 93 coal miners, who were killed in a methane gas explosion in the Hokutan Coal Mine in 1981. The signboard on the Shinto shrine reads "93" in Japanese. This is an example of archived memories in the form of a religious structure for events that happened in the coal mine. Interestingly, a sign at the site indicates that the shrine was not directly related to coal miners killed in the methane gas explosion. Some experts familiar with the coal mining scene in Hokkaido indicated that the shrine was a prop made by a popular culture group for photography. Beside this popular cultural representation of a shrine, the rusted iron grills that blocked the entry point of the Hokutan Coal Mine are now semi-covered by shrubs. Nearby, there is a tombstone with inscription to appease the souls of the 93 miners who died in the accident at Hokutan.

[64] Kushiro City Museum, "Kushiro City Museum" in the City of Kushiro website [downloaded on 1 Jan 2017], available at https://www.city.kushiro.lg.jp/common/000052727.pdf

[65] Best Friends Hokkaido (BFH), "Tracing Memories of the Coal Mining Boom: Messages Left in Sorachi" in the BFH website [downloaded on 1 Jan 2017], available at http://www.bfh.jp/en/lifestyle/tracing-memories-of-the-coal-mining-boom-messages-left-in-sorachi/

[66] Best Friends Hokkaido (BFH), "Tracing Memories of the Coal Mining Boom: Messages Left in Sorachi" in the BFH website [downloaded on 1 Jan 2017], available at http://www.bfh.jp/en/lifestyle/tracing-memories-of-the-coal-mining-boom-messages-left-in-sorachi/

Fig. 4.6 A rusted stand near Hokutan Coal Mine with written slogans articulating hopes for safe and secure output and production of coal. This is a relic of the mothballed coal mines in Yubari. (Photo taken by the author on 8 July 2016)

Besides physical structures, some former coal mining towns have turned to organizing events and non-traditional tourism to generate local community revenue. The gentrification and service/retail conversion of Yubari and its industries was officially known by the policy of "tanko kara kanko e (mining to tourism)" in arresting the pace of depopulation after the coal industry was phased out.[67] The heady construction of entertainment and touristy

[67] Seaton, Philip, "Depopulation and Financial Collapse in Yūbari: Market Forces, Administrative Folly, or a Warning to Others?" dated Winter 2010 in the Social Science Japan Journal Vol. 13 No. 2 [downloaded on 1 Jan 2017] (UK: Oxford University Press),

areas resulted in the construction of the Historical Coal Mining Museum and Amusement Park (1983), which added to the existing list of attractions, such as Mt. Reisui Ski Resort.[68] Today, the gentrified parts of Yubari comprise the site of a well-known international film festival and musk melons. Because of its quaint "Northern Exposure" kind of atmosphere, Yubari is also drawing in non-traditional tourists, highlighting the potential of the region for non-traditional tourism, marketing local community attractions, and sustainable ecotourism. The Geopark in Mikasa, for example, is well visited by local school children as well as by other Japanese visitors keen to study the park's natural resources, including professional palaeontologists studying ammonites whose fossils were deposited in the region when it was still an ocean floor around 66–100 million years ago.[69]

An example of the unconventional tourists visiting Yubari is the *bozo-zoku* or motorcycle convoy, with their heavy and powerful motorbikes riding unrestrained through Hokkaido's vast space. They are really motorbike enthusiasts role-playing a motorcycle tribe lifestyle. They are one example of non-traditional tourists and visitors to the wide open spaces of Hokkaido. The ageing population and hollowing out of former coal mine towns and agricultural areas meant that Bibai had wide spaces for outdoor activities. During his fieldwork trip on 8 July 2016, the author saw a number of bikers riding machines like Harley Davidson along Hokkaido's highways on the weekend. Entire convoys could be seen. Bikers were therefore an important stakeholder in accessing and using the natural landscape found in Bibai. The author came across the convoy that made a pit stop at the side of the highway. Most riders were almost like cosplayers, wearing authentic leather gear that mirrored American bikers. They portrayed a gritty style that was deliberately reconstructed for authenticity.

Retro movie posters greet visitors to Yubari, persistent features of the city's iconic film festival. Both Hollywood and local films are featured in this festival. Yubari is a place for film buff enthusiasts, signified by a hand-drawn retro poster featuring the legendary Charles Bronson (who later

available at https://www.jstor.org/stable/pdf/40961265.pdf?refreqid=excelsior:db749056af5c857ae4ec24e37b1de1cf, p. 228.

[68] Yanagida, Ryozo, "Yubari: Revitalization of coal mining city" dated September 2006 in the Meiji Gakuin University website [downloaded on 1 Jan 2017], available at http://www.meijigakuin.ac.jp/~hattori/shrinkingcity/page/JAPAN/yuubari-slimmed-103208k.pdf

[69] Taniguchi, Masaharu, "Mikasa Geopark Exploring the inner workings of our planet Earth" in the Hokkaido Magazine Kai [downloaded on 1 Jan 2017], available at http://kai-hokkaido.com/en/feature_vol32_mikasa/

became the chairperson of the powerful pro-gun lobby group National Rifle Association) from the 1970 movie *Violent City*. Incidentally, Bronson himself was a former coal miner from Pennsylvania. Images of James Dean who exemplified a rebel without a cause are featured in a hand-drawn movie poster at Yubari. These posters seem to symbolize nostalgia and a revival of sentimental memories of silver-screen icons. Bruce Willis' iconic *Die Hard* movie is featured in a nostalgic hand-drawn cinematic poster. The faded appearances give the town a nostalgic Showa-era feeling. Along the slopes, there are cut-out displays against which visitors can position their faces to pose for photographs.

Besides Hollywood film features, the city also features a sleek well-designed Bibai museum building. It indicates the potential for Bibai to be a major coal historical education centre, particularly in northern Japan. This potential local tourism site is a relatively well-developed depository of local knowledge and experience based on the coal industry. There are potential stakeholders in this project, including schools in the region or former coal mining towns that are interested in equipping their students with historical awareness of the coal mining industry and any Japanese visitors curious about Bibai history as well as its coal mines. Contents of the museum can be divided into three categories. To attract foreign visitors, the museum would have to include English- and foreign-language captions and detailed signboards.

In Kushiro (eastern Hokkaido), the local community has turned local climatic conditions of short summers and foggy weather to their advantage by celebrating them (instead of succumbing to the elements) through a *matsuri* or a festival lasting three days in July that features a laser show and a *hanabi daikai* fireworks event.[70] Besides traditional festivals, in July 1992, Bibai created an artistic contemporary space out of a secondary forest at a former coal mine and a defunct school structure (the Bibai public Sakae elementary school formerly educating 1200 miners' children from a Mitsubishi coal mine in East Bibai) at the Arte Piazza Bibai, adorned with 40 sculptures crafted by internationally famous Kan Yasuda sitting on 70,000 square metres of property.[71] The ground level of the former school is currently a kindergarten while the upper floor is a sculpture gallery in a

[70] Mizu no Kamuy Tourism Area, "Kushiro City 2 Festivals in Kushiro" in the Mizu no Kamuy Tourism Area website [downloaded on 1 Jan 2017], available at http://kam-kankouken.jp/tourism/en/kushiro/special/459.html

[71] Hokkaido Bibai City Portal site, "Arte Piazza Bibai" dated 2008 in the PiPa Hokkaido Bibai City Portal website [downloaded on 1 Jan 2017], available at https://www.pipaoi.jp/fl/en/artepiazzabibai/

nostalgic old-school classroom setting and musical performances (classical, jazz, etc.) are held in the gentrified gymnasium of the former school building and in the outdoor marble platform.[72] The Arte Piazza Bibai was specifically create to curate memories of the coal mines through the local authorities' invitation for Italy-based Bibai-er Kan Yasuda (who coincidentally was searching for a Japanese studio) to sculpt the Yamano Hi (Monument of the Mine) in Garo Family Park in 1980 and set up the artistic space of the Piazza at the wooden building of the former Sakae elementary school, which also shut down in 1980.[73] While Bibai may have lost its famous ballet troupe, its cultural heritage and status are making a big comeback with the Arte Piazza.

Agriculture and Food Industries

Another thriving industry is agriculture as farmers cultivated crops and set up wineries in former coal mining regions. For example, the Yamazaki family who owns Yamazaki Winery opined that the grade and soil composition of Mikasa's Mount Tappu are suitable for growing premium grapes used for wine-making (especially for making pinot noir and chardonnay wines).[74] Because of Mikasa and other wineries in Sorachi, wine manufacturing which has grown up organically has become a tourist attraction by itself, with group activities such as the "Sorachi Wine Picnic" in Iwamizawa.[75] In the case of Bibai, a potential Bibai attraction is the food that the city serves, including Hokkaido-style miso ramen and chicken rice steamed in cast-iron pots. Miso ramen is often associated with Hokkaido, Tokyo-ites seem to prefer *shoyuu* (soy sauce ramen), while the south in Kyushu is well known for its pork broth ramen. Besides fermented beverages and cooked food, other high value-added premium products include fruits.

[72] Hokkaido Bibai City Portal site, "Arte Piazza Bibai" dated 2008 in the PiPa Hokkaido Bibai City Portal website [downloaded on 1 Jan 2017], available at https://www.pipaoi.jp/fl/en/artepiazzabibai/

[73] Hokkaido International Center, Japan International Cooperation Agency, Regional and Industrial Promotion Study in Hokkaido Study Report (Japan: Hokkaido International Center, Japan International Cooperation Agency, Hamanasu Foundation), 2015, p. 17.

[74] Taniguchi, Masaharu, "Mikasa Geopark Exploring the inner workings of our planet Earth" in the Hokkaido Magazine Kai [downloaded on 1 Jan 2017], available at http://kai-hokkaido.com/en/feature_vol32_mikasa/

[75] Best Friends Hokkaido (BFH), "Tracing Memories of the Coal Mining Boom: Messages Left in Sorachi" in the BFH website [downloaded on 1 Jan 2017], available at http://www.bfh.jp/en/lifestyle/tracing-memories-of-the-coal-mining-boom-messages-left-in-sorachi/

The origins of the Yubari melon started with its consumption by Yubari-based Westerners working as foreign experts in the 1930s; therefore, only a few select farms catered to this niche market, but by the 1990s, local Japanese grew to appreciate the fruit and it quickly became one of Yubari's main farm produce.[76] In Yubari, the premium musk melon products continue to set records for the price ever paid to fruits, the first pair harvested in 2014 was sold for 15,000 pounds.[77] The Yubari melon is a ubiquitous symbol in the city. The products are so famous that they come with their own mascots in the form of both cute and fearsome-looking bears. Yubari's cute (kawaii) melon bear continues to ride on the popularity of the musk melons and its merchandises was ubiquitously retailed in souvenir shops throughout Yubari. The Yubari mascot is a bear wearing a musk melon cap. The black and white bears look majestic bearing their impressive canines. The local farming community also organizes an annual "all you can eat" melon buffet. In 2016, it started on June 10. There were also cardboard displays in the hotel where the author stayed that had face cut-outs where kids can pose through holes for photo-taking. Premium sodas were also sold in Yubari with melon flavours that come with 1% juice content. Premium melons often came with their own stickers bearing quality grading, identification of originating farm, and its own branding.

Bibai was another former coal mining area that historically had an active agricultural sector. Bibai City (central Sorachi) was founded by Isojiro Fukushima, a settler at Bibai River who became a ferryman in 1886, and as the settlement grew in size into a town, the farming industry in Bibai cultivated wheat and rice.[78] Bibai has also gone into green tourism based on farmers' markets, by retailing food produce with high levels of food safety and by offering fieldtrips for students to have practical experiences in handling farm produce and on-the-job work at farms every spring and

[76] Yanagida, Ryozo, "Yubari: Revitalization of coal mining city" dated September 2006 in the Meiji Gakuin University website [downloaded on 1 Jan 2017], available at http://www.meijigakuin.ac.jp/~hattori/shrinkingcity/page/JAPAN/yuubari-slimmed-103208k.pdf

[77] The Guardian, "Yubari, Japan: a city learns how to die" dated 15 August 2014 in The Guardian [downloaded on 1 Jan 2017], available at https://www.theguardian.com/cities/2014/aug/15/yubari-japan-city-learns-die-lost-population-detroit

[78] Hokkaido International Center, Japan International Cooperation Agency, Regional and Industrial Promotion Study in Hokkaido Study Report (Japan: Hokkaido International Center, Japan International Cooperation Agency, Hamanasu Foundation), 2015, p. 16.

autumn seasons.[79] During his fieldtrip on 8 July 2016, the author passed by a roadside store selling Yubari musk melons. The store also offers delivery to any location in Japan. The well-known Yubari musk melon with its characteristic fragrance and sweet and juicy texture was sampled by the author. Overshadowed by the well-known Yubari musk melon, Yubari watermelons are also well-regarded by locals when they are harvested in summer. Japanese people eat watermelons to enjoy the refreshing cooling effect in the heat of summer. The external corded texture of a hand-picked high-quality musk melon carries a Yubari quality assurance sticker, packed gingerly into a box for shipping long distances.

In areas where agriculture and traditional fishing and farming thrive, local community life and activities are rich with communitarian events like festivals and rituals. Kushiro is particularly colourful and rich in such activities. In Kushiro, a thriving fishing industry leads to the Kushiro Port Matsuri (Festival), a tradition started in 1948 and which became Kushiro's largest event over three days in the first week of August, involving floats designed to look like fishing boats as a means to worship local gods and give thanks for bountiful harvests of fish in Kushiro.[80] The Kushiro Tairyo Donpaku is another annual agricultural harvest and fishing *matsuri* (festival) lasting three days from the start of September which includes eating, drinking, and viewing boisterous fireworks to commemorate the autumn harvests and catches.[81]

Another festival or *matsuri* related to agricultural produce is found in Mikasa's annual Ume (Japanese plum blossom) Festival happening every May, and the main event features a competition that involves long-range spitting of locally pickled plums into a pit.[82] In terms of a *matsuri* without an agricultural theme, the annual One-Million Kite Festival in Bibai is organized at the city's Sky Port Bibai airport, featuring a competition

[79] Hokkaido International Center, Japan International Cooperation Agency, Regional and Industrial Promotion Study in Hokkaido Study Report (Japan: Hokkaido International Center, Japan International Cooperation Agency, Hamanasu Foundation), 2015, p. 17.

[80] Mizu no Kamuy Tourism Area, "Kushiro City 2 Festivals in Kushiro" in the Mizu no Kamuy Tourism Area website [downloaded on 1 Jan 2017], available at http://kam-kankouken.jp/tourism/en/kushiro/special/459.html

[81] Mizu no Kamuy Tourism Area, "Kushiro City 2 Festivals in Kushiro" in the Mizu no Kamuy Tourism Area website [downloaded on 1 Jan 2017], available at http://kam-kankouken.jp/tourism/en/kushiro/special/459.html

[82] Mikasa City, "The City of Mikasa: a great place where you can enjoy!!" in the Mikasa City website [downloaded on 1 Jan 2017], available at http://www.city.mikasa.hokkaido.jp/hotnews/detail/00001591.html

involving a variety of kites in an assortment of sizes made by the local community alongside shows put up by diehard kite enthusiasts.[83] *Matsuris* are often boisterous and effective in attracting tourists for a memorable visit as they experience collective participation in a ritual and feel the electric atmosphere of a town coming together to partake in an important community activity. The intangible *matsuri* cultural experience reveals much about how communities are organized and the interpersonal relations between participants.

East Hokkaido Fieldwork 2017

The bulk of the author's fieldwork in East Hokkaido was focused on Kushiro. The Kushiro Coal Mine operated by the KCM firm was the first Japanese coal mine facility that the author visited which is still operational. This was the smaller-scale contemporary version of the coal mining company with limited coal production output after the coal mine closed. Therefore, it was perhaps the most important case study, since the author was able to visualize workers at work in the mine and its daily operations. The author's visit to Bibai in West Hokkaido was brief. The facilities for coal mining workers are still standing and in use. Local pride in the industry is one major draw of the industry that manages to attract younger coal miners through Hello Work initiatives.

The miner quarters are still occupied by current KCM (Kushiro Coal Mine) staff members. Working within the proximity of their homes was also a draw for local coal miners. Hello Work organizations and offices provide free-of-charge administrative support for career building and job searches funded by the Japanese government as a bid to match job-seekers with the appropriate employment opportunities. KCM buildings are simple constructions with riveted cast-iron skeletal structures and zinc-roof shelters. Inside KCM, lockers are provided for miners and there are hatches where miners can deposit their dusty uniforms after work for laundry services. Special smoking areas are marked out and designated with fire alarm facilities, ensuring early warning to prevent fire incidents. The coal mine facility is peppered with reminders to take precautions and practise the concept of safety first. The five cardinal rules of safety in KCM read (translated from the Japanese language):

[83] Hokkaido Bibai City Portal site, "Arte Piazza Bibai" dated 2008 in the PiPa Hokkaido Bibai City Portal website [downloaded on 1 Jan 2017], available at https://www.pipaoi.jp/fl/en/artepiazzabibai/

1. Protect oneself.
2. Protect others.
3. Follow through decisions.
4. Avoid doing anything that one is unsure of.
5. Ask if unsure.

Beyond the lockers, one arrives at the helmet LED lamp headlight charging facilities, where there are light indicators to show whether light bulbs are fully charged or still charging. The light bulbs are state of the art, which indicated that the company gives maximum priority to the safety of their personnel through equipment investments. Given the fact that the Taiheiyo mine in Kushiro is an underwater coal mine, the reliability of headlights is very important. One could also tell whether a particular individual has gone to work from the absence or presence of his LED lamp in the charging facility. The helmet LED lamp headlight charging facilities also distinguished between "OB" (old boys, mean veteran workers, or upper-study seniors) versus "Trainees" (*Kensyuu*). This may suggest that there is some on-the-job training (OJT) in the facility, with skills passed down from *senpai* (upper study) to *kohai* (lower study). The workspace is also carefully rationalized, with equipment under repair placed in slots that are marked out with "Under Repair" signs.

There are also Kanban signs with rings to loop around objects or equipment that are not functioning well. Posters are pinned up on old safety chalkboards, with various notices on issues important and related to miners' safety. Old zinc roofs on some of the facilities are replaced with plastic-roof coverings, signifying reuse of old facilities to minimize costs but replacing components and parts where necessary. Fire marshals carefully check extinguisher equipment, with chalk boards indicating dates of inspection. There are careful and conscientious follow-ups on all safety operations, down to the most basic procedures. The schedule for rail transport to the undersea mines is displayed on a monitor, with time intervals clearly indicated. When the author arrived at the mine, Identification No. 201 coal miner had just entered the extraction area with the time of entry accurately indicated. Garbage and refuse are carefully discarded in corners marked out for such items. A banner proudly indicated that there had been no accidents/incidents for 1000 days at the time of the author's visit.

At the Kushiro Coal Mine, as the coal mine workers are about to descend into the depths of the coal mine, a sign warns them about the sudden drop in air pressure. Kushiro mines are located under the sea and extend into the coastal reaches of the Pacific Ocean. Railway tracks, freight cars, and pumps are operated by coal mines underground. The corridor leading to the mines was also well-lit with florescent daylight. In terms of spatial layout, the office space in which the office workers worked was uncluttered and neat, with landlines provided for communication and simple in-out trays. Loose sheets of paper were absent, as they were pinned down on the desk by transparent plastic coverings. There were no large pieces of equipment nor digital equipment to clutter the area. The mine workers were also able to rest at ease and many of them engaged in conversations where they exchanged information about their experiences inside the mine. For foreign understudy workers, the company was meticulous in translating safety instructions into their native languages to prevent misunderstandings. One such notice at the office indicated the following points in the Chinese language (translated from Chinese into English by the author):

Coal miners entering the mines must know the following:

1. Before exploration, one must first check his own equipment.
2. One must bring gasification safety equipment for rescue.
3. Observe safety regulations, to facilitate work in the mines, wear a safety helmet, uniform, and safety shoes.
4. Do not bring unneeded fire-related items to the workplace.
5. Do not bring along dangerous flammable items.
6. Do not casually take work equipment out of the mine.
7. Maintain good health constantly and enter the mine with a positive attitude.
8. Do not consume alcohol and then enter the mine.
9. Do not run when entering or exiting the mine, do not squeeze when moving, move with order.

If anyone does not observe the above points, company officials can reject their entry into the mine.

The same notice is also translated into the Vietnamese language. Upon leaving the mines, some coal mine workers left their equipment in open spaces on long benches as they dislodge their equipment before moving to the rest area. The spaces are rationalized, open, and uncluttered, just like the rest of the mine entry space. Smoking areas were clearly marked out with ashtrays and fire extinguishing sand buckets. Signs at the rest area urged coal mines to practice self-initiatives in observing safety, constant self-inspection of safety standards, maintain honesty, sincerity, and integrity as well as following rules. After dislodging all equipment, workers proceed to the common bath area where there is a relatively deep pool for them to soak and wash away coal soot. There were special taps that reach upwards, designed for coal mine workers to wash their eyes off any coal soot. They then take a shower and soap off before entering into an enticing *ofuro* hot bath. It was a simple spacious facility but probably the most welcomed part of coal mine exit procedure. A digital weighing scale that also has height readings enables workers to track their vital statistics for health. In the common bathroom, the *setsuden* or electricity saving campaign is implemented as workers are reminded to turn off the lights when they are not in use. Some workers may enjoy a smoke break after the soothing bath, so a corner is marked out for this purpose.

Another welcoming sight is the canteen, which operates from 11 am to 2.30 pm. The author smelt delicious wafts of Japanese curry when he passed by the canteen. Some workers prefer to bring their own bento boxes to eat at the canteen. Recycling is practised strictly here, as used tempura oil is deposited into a box for recycling purposes. Rice and noodles have different receptacles so that food-serving procedures are rationalized and orderly, as queues are split into two based on preferences to speed up the process. Like the rest of the company, the canteen is spacious and uncluttered, with simple tables accompanied by toothpick holders, soy sauce bottles, and napkins. Near the canteen, the author was touched by the display of a pair of origami cranes artistically folded to resemble the local Kushiro Tancho cranes. The origami cranes were facing each other and separated by a yellow paper flower. Decoration was simple and basic and conveyed local community's affection for its wildlife mascot (Fig. 4.7).

In terms of types of freight cars used, the KCM models included troughs placed on wheeled carts. They rested outside large warehouse-like facilities and a conveyor belt as well as pickup trucks for transporting light equipment around the mine. A weather vane was seen in the facility. Another KCM coal mining facility proximate to the mine was marked out for retailing and sorting out the extracted resources. After undergoing sorting and processing, the coal bits are then carried on a long conveyor belt to its final destination. The author and his team members traced the route using a vehicle. The conveyor belt is elevated and located above ground and part of its terminal point is located at a railway system owned by the company. This railway system runs 4 kilometres to the coastal port facilities, where coal is shipped off to its final destination. When a new power station is constructed in the area, the coal resource will serve the

Fig. 4.7 The main entrance of Kushiro Coal Mine. (Photo taken by the author in July 2017)

electricity power needs of that nearby station. The distance travelled will become shorter, since the station will be located near the KCM facilities, lowering the cost price for KCM coal. Feasibility studies will then be carried out to see if this diesel-powered railway system is optimal for transporting KCM coal (Figs. 4.8, 4.9, 4.10, 4.11, 4.12, 4.13, 4.14, 4.15, 4.16, and 4.17).

Unlike Kushiro's coal mine, Urahoro's former coal mine is left in ruins. To preserve memories of this site, signboards are used to mark former facilities found in this mine. These signboards pinpoint the former locations of mining quarters, sports facility, secondary school, bazaars, shops, primary school, public bath, hospital, and so on. During the author's fieldwork in July 2017, he was able to visit the remnants of a coal miner dormitory that used to house about 3000 individuals and the remains of a

Fig. 4.8 The sheltered conveyor belt system that processes loaded excavated coal and then transports them to the terminal train station for depositing into freight cars and then shipped off to the port. (Photo taken by the author in July 2017)

former archway at the entrance of the tunnel at the Urahoro coal mine site. In post–coal mine closure of Onbetsu as well as Kushiro, which still has a small-scale coal mining industry, the author spotted countless numbers of solar farms and use of solar panels throughout the city. The large roadside sign of the Onbetsu solar farm is visible from the highway. Solar panels are a renewable form of energy that give off very little polluting by-products. Given that the energy is used locally, energy loss is minimized. But solar energy alone cannot completely replace coal energy.

The author came across a large foreign coal inventory facility where Australian coal was stocked up for the local industries, of which the three largest are paper mills, agriculture, and fishing. Paper mills are the largest users of coal energy, and in Kushiro City itself, there are two large paper mill facilities with their own privately managed coal-powered plants for generating electricity. Here, foreign-imported coal is scooped up by

Fig. 4.9 A mothballed mechanized cutter displayed outside the KCM museum at the former site of the Skyland resort. (Photo taken by the author in July 2017)

excavators for use in power plants that are owned by paper mills. A new coal facility is being built that can capitalize on using the still-operational Kushiro Coal Mine (KCM) coal output. Because Kushiro coal is used within the city, transportation costs can be minimized and, correspondingly, energy wastage through coal transportation can also be reduced. When the new coal power plant is completed, stakeholders including KCM will have to decide if they wish to continue to rely on prevailing coal railway tracks or trucks plying the highways for deliveries.

The presence of solar power contributed to the pristine environment in former coal mining towns in Hokkaido, which is a major attraction for visitors to the region. One of the best ways to travel around this vast but sparsely populated territory is to rent a car. Water resources are major sites of tourism in Kushiro. Lake Mashu, Lake Kussharo, and Lake Akan are the three major tourist water bodies in Kushiro highly recommended by the local tourism authorities. Kushiro's frequent misty weather sometimes

Fig. 4.10 The terminal end of the coal conveyor belt at the KCM railway station. (Photo taken by the author in July 2017)

gives these lakes a fairy-tale like appearance. The image of Kushiro shrouded in mist was personally witnessed by the author during his July 2017 fieldtrip. Lake Akan is also home to a curious form of algae known as *marimo*, green algae balls found in the lake that can be bottled up and taken home, within local regulatory guidelines. Winter activities like skating can also take place here in winter when the water body freezes over (Fig. 4.18).

A major tourist attraction here is canoeing down Kushiro River. To experience this, the author participated in a canoeing expedition on 15 July 2017 through the former Ainu (indigenous natives of Hokkaido when it was known as "Ezo") territory. For visitors who are interested in learning more about Ainu culture, they can first visit a well-curated Kushiro museum. Even at the airport, basic Ainu culture displays, including the indigenous people's textiles (known as *ciciri*), salmon skin shoes (*cepker*), *pasuyop* (chopstick holders), *sapanpe* crown, *matanpus* headband,

Fig. 4.11 The D801 is a privately owned KCM coal transport train system that is diesel powered. It is a reliable workhorse. This train is built under licence from General Electric Company, US. Most, if not all, of Hokkaido's railway systems are modelled after US systems, while other parts of Japan adapted the continental European systems like the German railway model. In the licenced agreement, the local Japanese partner Nippon Sharyo worked with the General Electric Company to manufacture Model U10B Diesel Electric Locomotive weighing 55 tons in August 1970. Besides Model U10B D801, the author also spotted an older model built in 1966. (Photo taken by the author in July 2017)

ceninyupe pillow, and so on, are displayed. This airport display is probably any Kushiro visitor's first exposure to Ainu culture upon stepping into the city. For the adventurous visitors, Kushiro offers great potential for outdoor activities, including kayaking and canoeing. There are two well-known areas for such water-sports activities, including the inland lakes as well as the southeast Pacific coasts. The author participated in an expedition through the internal canoe down Kushiro River for experiential learning. In a bid to preserve the integrity of nature, the sign at the canoeing grounds of Kushiro Shitsugen National Park warned visitors not to do the following acts:

Fig. 4.12 KCM observes traffic safety, with junctions carefully barricaded to block cars from crossing when the railway train passes through. (Photo taken by the author in July 2017)

- Do not pick wildflowers or damage plants.
- No smoking while walking.
- Do not feed any wild animals.
- Care for nature.
- No pets allowed.
- Do not throw rubbish.

In the wilderness itself, the canoeing trip was pleasant and enjoyable, and the author was able to enjoy scenic views while spotting wildlife. Marshes and wetlands characterized most of the coastal areas around the Kushiro River. The canoeing trip is especially appealing to birdwatchers. In this trip alone, the author photographed an Ezo deer, wild ducks (probably a mallard), hawks (probably a black kite), karasu crows (one of these crows swooped in on the author's teammate when he was strolling in

Fig. 4.13 A wheeled car is placed across railway tracks that are unused to prevent any wandering pedestrians onto the mothballed tracks. The railway train currently moves between the terminal end of the KCM conveyor belt system and the port facilities only. (Photo taken by the author in July 2017)

Kushiro City on the morning of 16 July 2017), a dragonfly that accompanied the author in the canoe for part of the canoeing journey, and so on (Figs. 4.19, 4.20, 4.21, 4.22, 4.23, and 4.24). According to the Kushiro City Museum, the Hokkaido Ezo deer is a different species from the ones found in mainland Honshu. The museum caption for the Ezo deer explained: "The Kushiro area has a largest number of deer in Hokkaido because of its climate with a little snow-fall and its good food environment where their favourite bamboo grass grows thick." The *tancho* cranes are almost an unofficial ambassador for Kushiro and their likeness can be found almost everywhere in Kushiro City. Before canoeing, the author and other canoeists were informed about the presence of bears in the Kushiro River area using a bear-paw taxidermy artefact. For those visitors who are less inclined to do outdoor activities, another static display at the

Fig. 4.14 A KCM diesel refuelling facility for the train is located at the KCM railway station. (Photo taken by the author in July 2017)

Kushiro airport introduced the local zoo, which gives visitors the chance to view nature without wandering into the wilderness:

> Nearly 500 animals of 60 species live comfortably in the zoo's area of 47.8 hectares—one of the largest in Japan. Particularly popular with visitors is the Hokkaido Zone, which houses animals native to the island including the Blakiston's fish owl, the white-tailed eagle and the brown bear.

Besides the zoo, the Kushiro March Observatory is a platform for a view of the marsh without getting the feet wet. The Akan International Crane Centre is another facility for photographers and birdwatchers to spot and photograph migratory cranes feeding in Kushiro. The Kushiro City Museum was a treasure trove of artefacts related to the city. A large Taiheiyou coal bit sits at the entrance of the museum. The museum carefully curates and archives adult *tancho* cranes, taking care of their young

Fig. 4.15 The train station at the KCM railway facility. Here the station master carefully watches over the railway tracks. When the author visited the facility, the station master warned off a member of the public who had unintentionally wandered into the tracks outside the junction crossing. Safety is strictly observed in this facility. (Photo taken by the author in July 2017)

chicks. *Tancho* babies' delivery is celebrated with great fanfare in the city. *Tancho* cranes are dedicated parents who carry their babies on their back and feed the young with their beaks. They also fly to the middle of the marsh when bringing up their young, as that space is inaccessible to human and other predators.

In Kushiro City, the author also spotted what seemed like a Japanese wagtail and flocks of seagull at the Kushiro coastal area (Fig. 4.25). The author was unable to photograph a sea eagle at high altitude and *tancho* cranes parenting a young one near the Kushiro airport. During the author's arrival at the Kushiro airport, he spotted a sign for the *tancho* crane display introducing visitors to the Kushiro Japanese Crane Reserve that read:

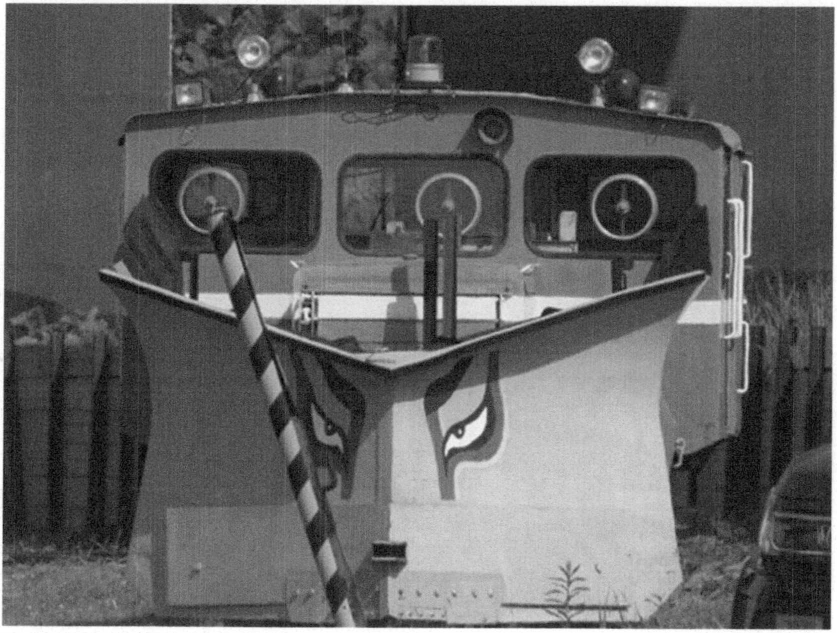

Fig. 4.16 A snow plough stands ready in case of heavy snowing at the railway tracks in the Hokkaido winter. (Photo taken by the author in July 2017)

> Japanese cranes living in a near-natural environment can be viewed from up close throughout the year. The facility is home to approximately 20 graceful cranes.

This is an important transformative role for the former coal mining city, as it is transformed from a site of resource extraction for heavy industrialization to an environmentally friendly nature reserve and nature observation platform. It also converges with Kushiro's new role as a pristine environment for ecotourism.

In the former Urahoro coal mine, Tokachi, the author photographed what looked like freshwater river *ayu* swimming in a brook near the abandoned coal mine site. The *ayu* (also known as sweetfish) are prized freshwater catches and taste good when salted and grilled over a slow fire. The presence of these fishes may be a reason why bears have come to inhabit this abandoned coal mine. A local administrator used cow bells to warn

Fig. 4.17 Each coal freight car weighs 12 tons and can carry 30 tons of coal load. Some of these freight cars were slightly stained by salt leaking out of the seams of the coal freight car. (Photo taken by the author in July 2017)

the bears that the author and his teammates were travelling into bear territory in a bid to prevent bear-human encounters. There were many species of insects including butterflies that thrived in the area. The author spotted what was probably a grey-veined white butterfly working on flower pollen near the former Urahoro mine, which bears are known to inhabit. Nature has reclaimed land formerly worked on by humans. Besides wildlife, another asset that Mother Nature left for Kushiro are the geothermal natural hot springs for the enjoyment of visitors and locals alike (Figs. 4.27, and 4.28).

For visitors interested in retail activities, they can visit Kushiro's Fisherman's Wharf for fresh barbequed and grilled seafood (Fig. 4.26). It was a good experience for the author to partake in local food anthropology. The Washo Fish Market is also a favourite stop for seafood ingredients' shopping. The Kushiro City Museum has an exhibit display caption

Fig. 4.18 The author's canoeing trip down Kushiro River, a major tourist draw for visitors, foreign tourists, and local adventurers. On the day of the author's canoeing expedition, his canoe came across a Taiwanese family on tour in the river, carrying out the same activities. (Photo taken by the author in July 2017)

to explain the prevalence of seafood culture in Kushiro: (1) "Kushiro is the fishery base for fish resources from the North Pacific and the Bering Sea" and (2) "[Maiwashi] are the species of a warm current though they migrate to the offshore of Kushiro in order to search for food. The Pacific Ocean off Kushiro is the good ground for growth of sardine." The museum signage on currents' condition and their relationship with the fishing industry reads:

> Offshore Kushiro is the place where the cold and warm currents meet. The cold current goes down to the south along the east coast of the Kuriles and meets the current from the Okhotsk and form the Kurlian Current. This current is abundant in nutritious salt and teems with zooplankton and phytoplankton which are food for the sea creatures.

Fig. 4.19 An Ezo deer at Kushiro River. (Photo taken by the author on 16 July 2017)

Fig. 4.20 A wild Japanese duck at Kushiro River. (Photo taken by the author on 16 July 2017)

Northern Japanese clams are tasty and fresh. Their prices have gone up due to the still-recovering clam supply chains in northeastern Japan (Tohoku area), which were disrupted by the Great East Japan Earthquake (also known as 311) and, resultantly, clam eggs were also washed away,

Fig. 4.21 A hawk, probably a black kite, at Kushiro River. (Photo taken by the author on 16 July 2017)

Fig. 4.22 Karasu crows at Kushiro River. (Photo taken by the author on 16 July 2017)

according to a veteran agricultural expert travelling with the author. Clams are an integral part of the local specialty *katte-don*, which consists of seafood ingredients like clams, raw fish, and fish roe. The author consumed clam sushi at a well-known local sushi chain. The texture is fatty, savoury, and chewy (Figs. 4.29, and 4.30).

Fig. 4.23 A double-bladed dragonfly at Kushiro River. (Photo taken by the author on 16 July 2017)

Fig. 4.24 A dried taxidermied bear paw used by the outdoor guide to indicate the presence of bears in the region around Kushiro River. (Photo taken by the author on 16 July 2017)

Fig. 4.25 Wagtails appear to be a common species of birds in the Kushiro region. The author spotted another white wagtail in the former coal mine of Onbetsu. (Photo taken by the author on 16 July 2017)

Fig. 4.26 The prevalence of local fresh squid (the author heard locals referring to it as white squid meat) meant that it was an important source of proteins for the Kushiro people and significant to their local culinary culture. (Photo taken by the author on 16 July 2017). Japanese pepper is added to mayonnaise as a dipping sauce for barbequed squid freshly caught from the proximate sea. Some locals like this hybrid combination as a kind of dipping sauce

Fig. 4.27 Freshwater river *ayu* swimming in a brook near the abandoned coal mine site. The *ayu* (also known as sweetfish) are prized freshwater catches and taste good when salted and grilled over a slow fire. The presence of these fishes may be a reason why bears have come to inhabit this abandoned coal mine. (Photo taken by the author on 16 July 2017)

Fig. 4.28 A local administrator used cow bells to warn the bears that the author and his teammates were travelling into bear territory in a bid to prevent bear-human encounters. Whistles are also used to warn the bears of humans approaching into their areas to prevent accidental encounters. (Photo taken by the author on 16 July 2017)

Fig. 4.29 The thriving fishing industry is evidenced by rows of fishing boat fleets parked at the Fishermen's Wharf. (Photo taken by the author on 16 July 2017)

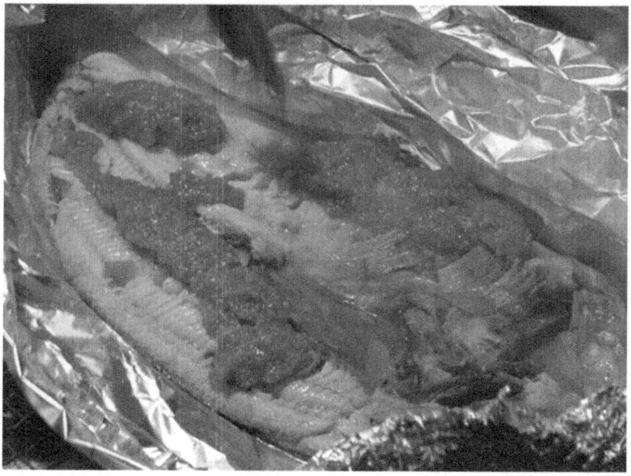

Fig. 4.30 Miso-laden fish caught locally is wrapped in aluminium foil and steamed/grilled inside the foil. Miso is a popular condiment used in local cooking. Another well-known Hokkaido-style miso-related dish is miso ramen, where miso is used instead of the usual pork broth in Kyushu-style tonkotsu ramen or shoyuu soy sauce–style ramen favoured in the Kanto area. (Photo taken by the author on 16 July 2017)

Fig. 4.31 A threshing machine at work in the Urahoro region, an example of mechanization of farm activities in this area. (Photo taken by the author in July 2017)

Besides the fishing industry, East Hokkaido also has a farming industry. All along East Hokkaido, the consciousness of environmental protection is strong. A sign at the entrance of Urahoro tells visitors and locals alike not to discard garbage and empty tin cans. A clean environment is crucial for agricultural development. According to an agricultural specialist travelling with the author, the region is specializing in the cultivation of beans, beet, potato, and wheat. According to that specialist, corn for feeding livestock was also cultivated here and so was hay, which was wrapped in bales for transport to farms to feed animals. Some farmers practise niche agriculture and farming, for example, raising lambs in this part of Japan (Figs. 4.31, and 4.32).

Community The religious diversity of the local community can also be seen through its religious facilities. These facilities include a church near the Kushiro train station, Buddhist temples like the Nipponzan Myohoji,

Fig. 4.32 Bales of hay are wrapped up for delivery as feed to livestock at farms. (Photo taken by the author in July 2017)

and so on. The church-like building is used by the community as a commercial wedding hall for Christian weddings. There is also a visible presence of new religions in Kushiro, adding to the diversity of religious beliefs in the city. The author also heard a narrative offered by a frequent visitor to this region that Shinto shrines in Hokkaido went through some degree of indigenization based on local interpretations of traditional Shinto architectures after the former Ainu-held Ezo territory was colonized by the Honshu *Wa* Japanese people. Shinto shrine designs were based on interpretations and understandings held by descendants of early pioneering migrants.

Other tangible heritage items include artefacts and documents related to the former Kushiro Coal Mine. In the Taiheiyou Tankou Shiryouushitsu (Taiheiyou Coal Mine Resource Room), the author examined mothballed display models of coal mine-related equipment such as the SD Coal Extractor, coal mining equipment with oxygen mask and other protective gear, walkie-talkie sets, and so on. Other artefacts on display included

Taiheiyou coal samples and textual materials like Kushiro maps drawn and edited by primary school students, copper casting of coal miners at work, and American manuals on longwall mining systems and coal output projection charts. A Kushiro Coal Mining vintage sign in this archive facility indicated that coal was a "Made in Kushiro" product made possible by 45 million years of underwater pressure. Other items displayed included photos of coal miners speaking to primary school kids, the former Japanese Prime Minister Nakasone Yasuhiro visiting the coal mines in 1984 (Showa 59th year) wearing coal mining gear, US visitors posing with sumo wrestlers, local *matsuri* festivals in coal mining towns, workers going to work in passenger rail cars in 1955 (Showa 30th year), and so on.

Some documents focused on historical facilities associated with the coal mining industry. Taiheiyo Skyland was perhaps the largest scale rest-and-relax (R&R) facility for coal miners as well as for outsiders (i.e. members of the public). A copy of the pamphlet displayed at the Taiheiyou Coal Mine Resource Room shows attractive women having a fun time at the pool with their kids, kids playing on water slides, a large dining room, and an event hall with long tables. Touting affordability and easy access to accommodation at the venue, the pamphlet featured Japanese tatami rooms (*washitsu*) and Western-style accommodations in a resort-like environment. The pamphlet promoted Skyland as an all-summer amusement park with a warm water pool and a "jungle"-themed hot spring. The theme park was dismantled following the decline of the coal mining industry, but remained an example of the cutting-edge entertainment complexes that Japanese coal miners enjoyed during their free time located right inside their working space.

The Skyland site is now a small but comprehensive KCM museum. At the entrance of the museum, a 6.048-ton block of coal greets the visitors. Elaborate models show how coal is mined under the Pacific coastal waters and then transported through rail in electric railway cars manufactured by Toshiba. The rail cars run through tunnels supported by mechanized shoulders and then the coal load that they carry is processed by KCM factories before they are transported through conveyor belts into sorters. The mechanized shoulders are mobile and can move in tandem with the excavators using connected movable pistons. For sections of the coal mine tunnels that are not supported by the mechanized shoulders, the seabed tunnels are supported by reinforced steel, which forms arches to hold up the weight of the seabed, and further supported by wooden panels that line the steel arches. Coal is cut using US company JOY's drum-cutters. The museum's basement is a mock-up of a coal tunnel completed with a gas leakage detection device. The museum indicated the company miners' reverence for the spiritual guardians of the coal mines by building Shinto shrines to make offerings to the local gods.

Research Centre Upon the closure of Japanese domestic coal mines, Kushiro's former coal miners and managers with their wealth of experience became a major source of technology and training for other East Asian coal miners, particularly those from Indonesia, China, and Vietnam which still have thriving coal mining industries. Within overland (continental) Southeast Asia, Vietnam is a major beneficiary of coal industry training offered by Japan. From 2001 onwards, Japan worked with Vietnam to train 1171 miners, staff members, and trainees in the coal and mineral mining industries, while thousands of others stayed on in Vietnam and took courses on underground drilling engineering, mechanical harvesting/salvage, mine design, and exploitation of tunnelling technology at the Vinacomin vocational school taught by Japanese specialists.[84]

Hokkaido was at the forefront of these training programmes and international exchanges. In 2002, the Ministry of Economy Trade and Industry (METI) transferred used equipment from Kushiro and other mining facilities in Japan to East Asian countries and trained their coal mining personnel (200 of them visit Kushiro and Ikeshima annually) to operate the used machinery, while dispatching Japanese engineers to Asian coal mining sites to set up the equipment.[85] Even today, Kushiro provides courses on management, mining, safety, equipment, and electric facilities for 160 foreign mining engineers and management staff members yearly from China, Vietnam, and Indonesia at the Kushiro Coal Mine and Coal Mining Technology Training Center, disseminating management know-how and technological transfers to the rest of Asia.[86]

[84] Minh, Nghia, Vietnam Chamber of Commerce and Industry, "Vietnam–Japan Strengthening Cooperation in Coal Mining" dated 22 April 2013 in the Vietnam Chamber of Commerce and Industry [downloaded on 1 Jan 2017], available at http://vccinews.com/news_detail.asp?news_id=28298

[85] Yoshimura, Yoshito and the Coal Division Agency of Natural Resources and Energy (Japan), "Closure of Japan's Last Large-Scale Coal Mine" dated 15 March 2002 in the Institute for International Studies and Training (IIST) World Forum website [downloaded on 1 Jan 2017], available at http://www.iist.or.jp/wf/magazine/0067/0067_E.html

[86] Japan Coal Energy Center (JCOAL), "Programs for Technology Transfer to Overseas" dated 2017 in the JCOAL website [downloaded on 1 Jan 2017], available at http://www.jcoal.or.jp/eng/work/05/

Concluding Section: Case Studies of Ageing Towns

Former coal mining towns in Hokkaido have also become local and international case studies for understanding the impact of a fast-ageing population. In the case of Bibai, it used to be a cultural hub with a ballet troupe and five theatres screening new Hollywood shows in the 1950s, but the town's population declined from 92,000 to 27,800, and dampened demand and consumption catalysed the shuttering down of retail shops in the city.[87] Bibai successfully held on to its approximate 30,000-strong population, who were coal miners in the past residing in east Bibai.[88] Yubari has also become a focus of international media attention to analyse the impact of ageing populations in Japanese towns and cities. *The Guardian* observed that this former coal town previously nicknamed as Japan's coal capital has seen its population evaporate by 90% in half a century, declining from 120,000 in 1960 to 21,000 when the last coal facility shut down in 1990, prompting coal miners to leave the town (the population declined further to 10,000 by 2014 and the remaining individuals consisted of mostly elderly folks).[89] In 1990, the final coal mine that was shut down was the Mitsubishi Sekitan Kogyo Minami Oyubari Mine.[90]

By 2016, Yubari's population went down further to 9000 through natural attrition and experienced hollowing out as younger residents moved to urban centres in other parts of Japan.[91] According to a *Bloomberg* report, the following measures were implemented by the authorities in response to the ageing and declining Yubari population: combining

[87] Fackler, Martin, "As Oil Prices Rise, Nations Revive Coal Mining" dated 22 May 2008 in The New York Times [downloaded on 1 Jan 2017], available at http://www.nytimes.com/2008/05/22/business/worldbusiness/22mines.html?mcubz=0

[88] Hokkaido Bibai City Portal site, "Arte Piazza Bibai" dated 2008 in the PiPa Hokkaido Bibai City Portal website [downloaded on 1 Jan 2017], available at https://www.pipaoi.jp/fl/en/artepiazzabibai/

[89] The Guardian, "Yubari, Japan: a city learns how to die" dated 15 August 2014 in The Guardian [downloaded on 1 Jan 2017], available at https://www.theguardian.com/cities/2014/aug/15/yubari-japan-city-learns-die-lost-population-detroit

[90] Hokkaido International Center, Japan International Cooperation Agency, Regional and Industrial Promotion Study in Hokkaido Study Report (Japan: Hokkaido International Center, Japan International Cooperation Agency, Hamanasu Foundation), 2015, p. 11.

[91] Mogi, Chikako and Yuki Hagiwara (Bloomberg), "Deserted Yubari tries creating new population nucleus in test case for Japan" dated 26 September 2016 in Japan Times website [downloaded on 1 Jan 2017], available at http://www.japantimes.co.jp/news/2016/09/26/business/deserted-yubari-tries-creating-new-population-nucleus-test-case-japan/#.WUcPReuGOpo

resources for educational institutions; cutting pay of the local bureaucracy; downgrading a hospital to clinic; ceasing state funding for redundant facilities like washrooms, pools, public parks, public transportation, and snow ploughing; and moving Yubari residents from the suburbs to apartments in the city centre.[92] All these measures act as living labs and test cases for other coal towns and cities with declining populations. Yubari was not the only city facing depopulation; in fact the entire Sorachi region faced the same challenges, as the region had 103 coal mines, which had an output of 15 million tons of coal in 1960 during the golden era of the domestic coal mining industry. Like all former coal mining towns, Mikasa's population declined from its peak of more than 60,000 denizens in 1962 to 9000 in the contemporary period.[93] Most remaining residents in former coal mining areas like Yubari and Bibai today consist of the elderly population. There were physical signs of a declining population during the author's fieldtrip on 8 July 2016. For example, the rusted shutters of a former shop in Yubari reflected population decline and ageing demographics in the city. Houses where coal miners and other Yubari residents used to stay are currently uninhabited.

Despite the hollowing-out effect of depopulation, schools in Yubari continue to preserve and impart historical knowledge of Yubari's coal mining past to their students. These schools also kept well-organized galleries or heritage corners of collected material artefacts that served as reminders to Yubari youths of the golden era of coal mining, which served as the genesis of their city's formation. Yubari's case study is observed keenly by international observers as an example of how cities are eased into declining gracefully in terms of lifestyle and elderly care and how physically reintegrating abandoned parts of the city back to nature serves as a kind of demonstrative showcase of nature slowly reclaiming formerly polluted lands. Community spirit is another positive feature of the Japanese spirit in coping with adversity. In 2007, after bankruptcy was declared, the whole nation of Japan rallied to cheer up the residents, showing again the

[92] Mogi, Chikako and Yuki Hagiwara (Bloomberg), "Deserted Yubari tries creating new population nucleus in test case for Japan" dated 26 September 2016 in Japan Times website [downloaded on 1 Jan 2017], available at http://www.japantimes.co.jp/news/2016/09/26/business/deserted-yubari-tries-creating-new-population-nucleus-test-case-japan/#.WUcPReuGOpo

[93] Taniguchi, Masaharu, "Mikasa Geopark Exploring the inner workings of our planet Earth" in the Hokkaido Magazine Kai [downloaded on 1 Jan 2017], available at http://kai-hokkaido.com/en/feature_vol32_mikasa/

undaunted *ganbaru* spirit that characterizes Japanese dogged persistence against adversities. An example of community support is indicated through sumo wrestlers voluntarily travelling to Yubari to cheer up its residents with wrestling demonstrations.[94]

Alternatively, for other observers, Yubari can also be a showcase of creative urban revival and demonstrate how the state, local community, and local industries can work together to create a new role for an ageing city. *The Guardian* noted that in 2014, new investments in a traditional Chinese medicine (TCM or *kanpouyaku* in Japanese) industry arrived in the city, creating a number of jobs, and *The Daily* also noted that infrastructure in Yubari is still pristinely intact, with functioning post offices, well-oiled fire engines, low-rent apartments, free snow-shovelling services, and functioning preventive healthcare.[95] Yubari can be an effective showcase model for other ageing cities in both ageing gracefully and providing a high quality of life for its residents still living in the city and possible future migrants. Japan's own media outfits are also active in studying the Yubari's decline as an important case study of post-Showa economic bubble period economy and reversal of ageing demographics in the city. The country's two largest dailies have based their top journalists in the city hall near the mayor's office to curate the city's decline through detailed on-site observations.[96]

Bibliography

Best Friends Hokkaido (BFH), "Tracing Memories of the Coal Mining Boom: Messages Left in Sorachi" in the BFH website [downloaded on 1 Jan 2017], available at http://www.bfh.jp/en/lifestyle/tracing-memories-of-the-coal-mining-boom-messages-left-in-sorachi/

Board, Jack, "Asia's Toughest Jobs: The fire and brimstone miners of Ijen Crater" dated 19 August 2017 in the Channelnewsasia website [downloaded on 19 August 2017], available at http://www.channelnewsasia.com/news/asiapacific/asia-s-toughest-jobs-the-fire-and-brimstone-miners-of-ijen-9134662

[94] Wagner, Wieland, "Saving Japanese Ghost Towns" dated 30 November 2007 in Spiegel Online [downloaded on 1 Jan 2017], available at http://www.spiegel.de/international/world/empty-theme-parks-lower-pay-saving-japanese-ghost-towns-a-520703.html

[95] The Guardian, "Yubari, Japan: a city learns how to die" dated 15 August 2014 in The Guardian [downloaded on 1 Jan 2017], available at https://www.theguardian.com/cities/2014/aug/15/yubari-japan-city-learns-die-lost-population-detroit

[96] Wagner, Wieland, "Saving Japanese Ghost Towns" dated 30 November 2007 in Spiegel Online [downloaded on 1 Jan 2017], available at http://www.spiegel.de/international/world/empty-theme-parks-lower-pay-saving-japanese-ghost-towns-a-520703.html

Campi, Alicia, "The New Great Game in Northeast Asia: Potential Impact of Energy Mineral Development in Mongolia on China, Russia, Japan, and Korea" dated 2013 in the Asia-Pacific Policy Papers Series [downloaded on 1 Jan 2017], (Washington DC: Edwin O. Reischauer Center for East Asian Studies), 2013, available at http://www.reischauercenter.org/en/wp-content/uploads/2013/06/RC-Monograph-2013-Campi_The-New-Great-Game-in-Northeast-Asia.pdf

Cochilco, Chilean Copper Commission, "Research on Mine Closure Policy No. 44" dated January 2002 in the International Institute for Environment and Development (IIED) website [downloaded on 1 Jan 2017], available at http://pubs.iied.org/pdfs/G00541.pdf

Culter, Suzanne, Managing Decline: Japan's Coal Industry Restructuring and Community Response (Hawaii: University of Hawaii Press), 1999

Curtis, Katina, "Look at Japan for clean coal power" dated 2 Feb 2017 in News Limited [downloaded on 2 Feb 2017], available at http://www.news.com.au/national/breaking-news/look-at-japan-for-clean-coal-power/news-story/784d9cb50af857bfa8cbc9f5bc90acc8

Fackler, Martin, "As Oil Prices Rise, Nations Revive Coal Mining" dated 22 May 2008 in The New York Times [downloaded on 1 Jan 2017], available at http://www.nytimes.com/2008/05/22/business/worldbusiness/22mines.html?mcubz=0

Fukushima, Atsushi, "Coal and Environmental Issues in Northeast Asia" dated March 2004 in The Institute of Energy Economics, Japan (IEEJ) website [downloaded on 1 Jan 2017], available at http://eneken.ieej.or.jp/en/data/pdf/242.pdf

Go Musee Japan, "Mikasa City Museum" in the Go Musee website [downloaded on 1 Jan 2017], available at http://go-musee.com/archives/general/221/

Guardian writers, "Undiscovered south-east Asia: remote towns and secret beaches" dated 10 May 2017 in The Guardian [downloaded on 10 May 2017], available at https://www.theguardian.com/travel/2017/may/10/undiscovered-asia-thailand-vietnam-cambodia-laos-bali-beaches-islands

Hewett, Jennifer, "Japan has good reason to burn coal for electricity" dated 3 Feb 2017 in The Australian Financial Review website [downloaded on 3 Feb 2017], available at http://www.afr.com/business/mining/coal/japan-has-good-reason-to-burn-coal-for-electricity-20170202-gu4csz

Hokkaido Bibai City Portal site, "Arte Piazza Bibai" dated 2008 in the PiPa Hokkaido Bibai City Portal website [downloaded on 1 Jan 2017], available at https://www.pipaoi.jp/fl/en/artepiazzabibai/

Hokkaido International Center, Japan International Cooperation Agency, Regional and Industrial Promotion Study in Hokkaido Study Report (Japan: Hokkaido International Center, Japan International Cooperation Agency, Hamanasu Foundation), 2015

Hokkaido Tourism Organization (Hokkaido Department of Policy Planning and Coordination, Bureau of Transportation Policy, Transportation Policy Planning Division), "The area surrounding Sapporo, Otaru and Noboribetsu (Central Hokkaido)" in the Hokkaido Tourism Organization website [downloaded on 1 Jan 2017], available at http://shinkansen-access.visit-hokkaido.jp/en/area/area04/spot0408.html

International Energy Agency (IEA), "Emergence of Southeast Asia as energy giant carries risks, opportunities" dated 8 October 2015 in the IEA website [downloaded on 1 Jan 2017], available at https://www.iea.org/newsroom/news/2015/october/emergence-of-southeast-asia-as-energy-giant-carries-risks-opportunities.html

Japan Coal Energy Center (JCOAL), "Programs for Technology Transfer to Overseas" dated 2017 in the JCOAL website [downloaded on 1 Jan 2017], available at http://www.jcoal.or.jp/eng/work/05/

Japan Times, "Nuke crisis a boon for struggling coal mines" dated 31 March 2012 in the Japan Times website [downloaded on 1 Jan 2017], available at http://www.japantimes.co.jp/news/2012/03/31/national/nuke-crisis-a-boon-for-struggling-coal-mines/#.WUXDzOuGOpo

Japan Times, "Tourists take part in a 'mountain rail bike' tour on the abandoned Kamioka Railway in Hida, Gifu" dated 23 May 2017 in the Japan Times website [downloaded on 23 May 2017], available at https://www.japantimes.co.jp/news/2017/05/23/national/efforts-afoot-turn-abandoned-rural-train-lines-tourist-attractions/#.WZo9oD4jGUk

Japan Travel and Tourism Association, "Bibai City Tomei Park" dated 2014 in The Must-see Japan website [downloaded on 1 Jan 2017], available at http://www.nihon-kankou.or.jp.e.wp.hp.transer.com/sakura/searchDetail.jsp?ken__=hokkaido&shi__=01215&id__=S0127

Jiji Press Photo, "Japan Coal Mine" dated 8 August 2014 in the Jiji Press Photo website [downloaded on 1 Jan 2017], available at http://jijiphoto.photoshelter.com/image/I0000WV746N7o9aU

Johnson, Keith, "Japan Bets on Nuclear, and Coal, for Future Power" dated 8 April 2015 in Foreign Policy website [downloaded on 1 Jan 2017], available at http://foreignpolicy.com/2015/04/08/japan-bets-on-nuclear-and-coal-for-future-power/

Kasuga, Yutaka, "Transfer and development of coal-mine technology in Hokkaido" dated 1982 in the Series: Japanese Experience of the UNU Human and Social Development Programme series; 48 [downloaded on 1 Jan 2017], available at https://d-arch.ide.go.jp/je_archive/pdf/workingpaper/je_unu48.pdf

Koppenborg, Florentine, "Will the silent comeback of coal threaten Japan's climate goals?" dated 6 April 2017 in the East Asia Forum website [downloaded on 6 April 2017], available at http://www.eastasiaforum.org/2017/04/06/will-the-silent-comeback-of-coal-threaten-japans-climate-goals/

Kushiro City Museum, "Kushiro City Museum" in the City of Kushiro website [downloaded on 1 Jan 2017], available at https://www.city.kushiro.lg.jp/common/000052727.pdf

Kusuhara, Naoki, "Some Problems of Unemployed Laborers in Coal Mining Industry" in The science reports of the Tohoku University 7th series, Geography Vol. 17 No. 1 (Japan: Tohoku University), 1968, pp. 105–116

Lim, Tai Wei, *Energy Transitions in Japan and China: Mine Closures, Rail Developments, and Energy Narratives* (UK: Palgrave Macmillan), 2016.

Lim, Tai Wei, "The State of Coal Industry in Northeast Asia" dated 10 August 2017 in IPP Review website [downloaded on 10 August 2017], available at http://ippreview.com/index.php/Blog/single/id/527.html

Lohr, Steve, "Japan's Big Energy Setback" dated 27 Oct 1981 in The New York Times website [downloaded on 1 Jan 2017], available at http://www.nytimes.com/1981/10/27/business/japan-s-big-energy-setback.html

Martin, Benjamin, "Japanese Mining Labor: The Miike Strike" dated Feb 1961 in Far Eastern Survey Vol. 30 No. 2 (American Council, Institute of Pacific Relations), 1959, pp. 26–30.

McHugh, Babs, "Japanese government planning to build 45 new coal fired power stations to diversify supply" dated 1 February 2017 in the ABC website [downloaded on 1 Feb 2017], available at http://www.abc.net.au/news/rural/2017-01-31/japan-coal-power-plants/8224302

Mealey, Rachel, "Japan spruiks 'highly efficient' coal-fired power plants as stop-gap measure to Australia's energy crisis" dated 1 June 2017 [downloaded on 1 June 2017], available at http://www.abc.net.au/news/2017-06-01/japan-spruiks-coal-fired-power-plants-to-australia/8577464

Mikasa City, "The City of Mikasa: a great place where you can enjoy!!" in the Mikasa City website [downloaded on 1 Jan 2017], available at http://www.city.mikasa.hokkaido.jp/hotnews/detail/00001591.html

Minh, Nghia, Vietnam Chamber of Commerce and Industry, "Vietnam – Japan Strengthening Cooperation in Coal Mining" dated 22 April 2013 in the Vietnam Chamber of Commerce and Industry [downloaded on 1 Jan 2017], available at http://vccinews.com/news_detail.asp?news_id=28298

Mizu no Kamuy Tourism Area, "Kushiro City 2 Festivals in Kushiro" in the Mizu no Kamuy Tourism Area website [downloaded on 1 Jan 2017], available at http://kam-kankouken.jp/tourism/en/kushiro/special/459.html

Mogi, Chikako and Yuki Hagiwara (Bloomberg), "Deserted Yubari tries creating new population nucleus in test case for Japan" dated 26 September 2016 in Japan Times website [downloaded on 1 Jan 2017], available at http://www.japantimes.co.jp/news/2016/09/26/business/deserted-yubari-tries-creating-new-population-nucleus-test-case-japan/#.WUcPReuGOpo

Myllyvirta, Lauri, "Southeast Asia is planning 400 new coal power plants—what does that mean?" dated 13 January 2017 in the Energydesk Greenpeace website [downloaded on 13 Jan 2017], available at http://energydesk.greenpeace.org/2017/01/13/southeast-asia-coal-plans-health-japan-indonesia/

NHK World, "Yubari: Mining the Nostalgia" dated 25 Oct 2016 in the NHK WORLD TV Journeys in Japan website [downloaded on 1 Jan 2017], available at https://www3.nhk.or.jp/nhkworld/en/tv/journeys/journey_20161025.html

Ong, Han Sean, "Sungai Lembing's evolution from a tin-mining hub to a tourist destination" dated 20 March 2014 in The Star Online [downloaded on 1 Jan 2017], available at http://www.thestar.com.my/news/community/2014/03/20/tunnels-from-a-glorious-past-sungai-lembings-evolution-from-a-tinmining-hub-to-a-tourist-destination/

Raslan, Karim, "An Indonesian Gold Hunter Dreams of Better Days – in tourism" dated 2 March 2017 in the South China Morning Post (SCMP) website [downloaded on 2 March 2017], available at http://www.scmp.com/week-asia/business/article/2075441/indonesian-gold-hunter-dreams-better-days-tourism

Seaton, Philip, "Depopulation and Financial Collapse in Yūbari: Market Forces, Administrative Folly, or a Warning to Others?" dated Winter 2010 in the Social Science Japan Journal Vol. 13 No. 2 [downloaded on 1 Jan 2017] (UK: Oxford University Press), available at https://www.jstor.org/stable/pdf/40961265.pdf?refreqid=excelsior:db749056af5c857ae4ec24e37b1de1cf, pp. 227–240

Shimazaki, Naoko, "Support for Workers Displaced in the Decline of the Japanese Coal Industry: Formal and Informal Support" available at http://www.jil.go.jp/english/JLR/documents/2015/JLR46_shimazaki.pdf, pp. 6–27

Streets, David G., "Northeast Asia coal trading center set up in Dalian" dated 1 July 1991 in the Glocom website [downloaded on 1 Jan 2017], available at http://www.glocom.ac.jp/column/1991/07/energy_and_acid_rain_projectio.html

Taek-Whan Han, "Northeast Asia Environmental Cooperation: Progress and Prospects" dated 23 September 1994 in the Nautilus Institute website [downloaded on 1 Jan 2017], available at http://nautilus.org/trade-and-environment/northeast-asia-environmental-cooperation-progress-and-prospects-3/

Takahara, Kanako, "Yubari tries downsizing to solve fiscal crisis" dated 1 March 2007 in the Japan Times [downloaded on 1 Jan 2017], available at http://www.japantimes.co.jp/news/2007/03/01/national/yubari-tries-downsizing-to-solve-fiscal-crisis/#.WUoHu-uGOUk

Taniguchi, Masaharu, "Mikasa Geopark Exploring the inner workings of our planet Earth" in the Hokkaido Magazine Kai [downloaded on 1 Jan 2017], available at http://kai-hokkaido.com/en/feature_vol32_mikasa/

The Guardian, "Yubari, Japan: a city learns how to die" dated 15 August 2014 in The Guardian [downloaded on 1 Jan 2017], available at https://www.theguardian.com/cities/2014/aug/15/yubari-japan-city-learns-die-lost-population-detroit

Toshitaka, Nagahiro, "Japan's Coal Cartels in the Interwar Period" dated 17 March 2014 in the World Business History Conference (WBHC) website [downloaded on 1 Jan 2017], available at http://www.worldbhc.org/files/full%20program/A6_B6_JapanscoalCartelsintheinterwarperiod_ToshitakaNagahiro_.pdf

UPI, "Operators of a Hokkaido coal mine today completed pumping" dated 26 Oct 1981 in the UPI website [downloaded on 1 Jan 2017], available at http://www.upi.com/Archives/1981/10/26/Operators-of-a-Hokkaido-coal-mine-today-completed-pumping/1687372920400/

U.S. Energy Information Administration, "Chapter 4. Coal" dated 11 May 2016 in International Energy Outlook 2016 [downloaded on 1 Jan 2017], available at https://www.eia.gov/outlooks/ieo/coal.php

Wagner, Wieland, "Saving Japanese Ghost Towns" dated 30 November 2007 in Spiegel Online [downloaded on 1 Jan 2017], available at http://www.spiegel.de/international/world/empty-theme-parks-lower-pay-saving-japanese-ghost-towns-a-520703.html

Walker, Beth, "Chinese Investment Stokes Global Coal Growth" dated 24 September 2016 in The Diplomat [downloaded on 1 Jan 2017], available at http://thediplomat.com/2016/09/chinese-investment-stokes-global-coal-growth/

Yanagida, Ryozo, "Yubari: Revitalization of coal mining city" dated September 2006 in the Meiji Gakuin University website [downloaded on 1 Jan 2017], available at http://www.meijigakuin.ac.jp/~hattori/shrinkingcity/page/JAPAN/yuubari-slimmed-103208k.pdf

Yoshimura, Yoshito and the Coal Division Agency of Natural Resources and Energy (Japan), "Closure of Japan's Last Large-Scale Coal Mine" dated 15 March 2002 in the Institute for International Studies and Training (IIST) World Forum website [downloaded on 1 Jan 2017], available at http://www.iist.or.jp/wf/magazine/0067/0067_E.html

CHAPTER 5

The Coal Industry in the Northeast Asian Context

Introduction

This chapter[1] briefly discusses some developments in the Northeast Asian coal industry that may have a regional impact on Japan and are worth mentioning here, including emerging sources of coal in Northeast Asia and trends in coal use by Northeast Asian economies. It also discusses common issues faced by all Northeast Asia coal consumers. Coal and its importance in Northeast Asia's energy mix is analysed through a comparative lens of coal trade in the East Asian region. There is a comparative element in this chapter contextualizing and contrasting contemporary Japanese coal activities with its neighbouring countries/economies. Ultimately, it provides a macro picture of the Northeast Asian coal industry outside the immediate confines of Japan. In addition, other than Japan, China is the world's largest consumer of coal energy.. Therefore it is given prominence in this chapter as a preview of future Northeast Asian coal use and also how its activities will influence coal demand, supply, investments, and power generation in the region.

[1] Part of this section appears in the limited circulation background brief—Lim, Tai Wei, EAI Background Brief No. 1329—COAL AND ITS IMPORTANCE IN CHINA'S ENERGY MIX (Singapore: National University of Singapore EAI), 2018 (28 Feb 2018)—and in my older and briefer blog article: Lim, Tai Wei, "The State of Coal Industry in Northeast Asia" dated 10 August 2017 in IPP Review website [downloaded on 10 August 2017], available at http://ippreview.com/index.php/Blog/single/id/527.html. It is also drawn from a limited circulation background brief written by the author that was unpublished at the point of writing this chapter.

T. W. Lim et al., *Coal Mining Communities and Gentrification in Japan*, https://doi.org/10.1007/978-981-13-7220-9_5

Historical Use of Coal in China

Japan, China, and South Korea are located in a region where, with the exception of Japan and South Korea, the rest of the region is coal-rich or has sizeable deposits of coal. In China, the regions of Shanxi, Inner Mongolia, and Shaanxi have been some of the largest suppliers of domestic Chinese coal. In terms of variety, the major genres of coal are lignite in east Siberia, Far East region of Russia, and Mongolia, while the rest of the Northeast Asia region has mostly bituminous and sub-bituminous coal with some anthracite.[2] Traditionally, heavy industrial demand for coal from sectors like steelmaking has been a major source of coal consumption in the Northeast Asian region.

Electricity generation is another guzzler of coal fuels in the same region. More recently, coal use in China reached 3.97 billion tons (physical tons) in 2015 and 3.78 billion tons in 2016.[3] Heavy coal use for electricity, heating, and industries also comes with its own set of challenges. The population density of Northeast Asian cities (especially first-tier cities in the region) meant that pollution was concentrated and dense in certain areas. This has led rising middle classes in those cities to express socio-political dissatisfaction, compelling regional and national governments to react and come up with policies to reduce pollution or to enforce penalties for anti-pollution laws.

In terms of current consumption, amongst the Organisation for Economic Co-operation and Development (OECD) Asian countries, Japan has the biggest coal consumption making up approximately 50% of the Asian region's total coal consumption (about 5 quadrillion Btu) in 2012, used almost equally for electricity generation and industrial applications with the demand from the latter expected to dip after 2020 due to the projected drop in steel production caused by Japan's ageing population and decreasing demand.[4] By comparison, South Korea's steel production

[2] Fukushima, Atsushi, "Coal and Environmental Issues in Northeast Asia" dated March 2004 in The Institute of Energy Economics, Japan (IEEJ) website [downloaded on 1 January 2017], available at http://eneken.ieej.or.jp/en/data/pdf/242.pdf, p. 9.

[3] Lin, Alvin, "Understanding China's New Mandatory 58% Coal Cap Target" dated 17 March 2017 in Natural Resources Defence Council (NRDC) website [downloaded on 17 March 2017], available at https://www.nrdc.org/experts/alvin-lin/understanding-chinas-new-mandatory-58-coal-cap-target

[4] US Energy Information Administration, "Chapter 4. Coal" dated 11 May 2016 in International Energy Outlook 2016 [downloaded on 1 Jan 2017], available at https://www.eia.gov/outlooks/ieo/coal.php

is likely to increase coal use while the electric power sector makes up 62% of total coal consumption in 2012, along with thermal power expansion.[5] Dwarfing both Japan and South Korea, China is the biggest user of coal globally consuming 76 quadrillion Btu of coal in 2012, making up 50% of global coal consumption with trends of slowing as China adjusts to the "new normal" in economic development and as the energy industry goes through fundamental transformations to deal with politically sensitive environmental problems.[6]

Power generation is the largest demand driver of coal use in Northeast Asia and many other parts of the world. In the case of Japan, it is not alone in coal energy use, and it is quite likely the regional coal industry will continue to persist for years to come even if Japan has mothballed much of its domestic coal industry. Japan and South Korea stand out as the only major Northeast Asian economies without a sizable domestic coal mining industry. Environmental pollution is also a major universal challenge that faces coal use in Northeast Asia and a common solution in the region has been to phase out smaller, less-regulated, family owned (even illegal) coal plants that use obsolete technologies. The result of closure of smaller-scale coal-powered plants is a shift towards the utilization of larger plants. For example, China is phasing out smaller-scale coal-fired plants in favour of large plants with capacities of 500,000 kW.[7]

There are regional environmental cooperation initiatives amongst the Northeast Asian states like China, Japan, Russia, Mongolia, and South Korea which have established formal diplomatic relations with each other after the Cold War. These countries realize the advantages of combining resources to tackle common transboundary and cross-border environmental problems and pollution (including those caused by coal use). In the early days of Northeast Asian cooperation, China had already expressed through a Senior Official's Meeting (SOM) in 1993 that Beijing is interested to interact/dialogue with Japan and South Korea on topics of

[5] US Energy Information Administration, "Chapter 4. Coal" dated 11 May 2016 in International Energy Outlook 2016 [downloaded on 1 Jan 2017], available at https://www.eia.gov/outlooks/ieo/coal.php

[6] US Energy Information Administration, "Chapter 4. Coal" dated 11 May 2016 in International Energy Outlook 2016 [downloaded on 1 Jan 2017], available at https://www.eia.gov/outlooks/ieo/coal.php

[7] Fukushima, Atsushi, "Coal and Environmental Issues in Northeast Asia" dated March 2004 in The Institute of Energy Economics, Japan (IEEJ) website [downloaded on 1 Jan 2017], available at http://eneken.ieej.or.jp/en/data/pdf/242.pdf, p. 27.

technology transfer and financial assistance mechanism building and proposed that the three countries embrace the "spirit of Agenda 21" to do a joint research project for clean coal combustion technology (CCCT).[8] The Chinese were less keen on joint research and monitoring systems for transboundary air pollution which are of greater interest to Korea and Japan.[9] However, cooperative initiatives are often affected by geopolitical rivalries, historical memories, and territorial disputes.

Working on common environmental problems including those caused by the use of coal fuel is useful for strengthening interdependence amongst the Northeast Asian states since most of them are either highly industrialized (like Japan and South Korea) or are large emerging economies with maturing growth (e.g. China). Cooperation in such areas was however mired in political differences. Rapid industrial growth (and the accompanying urbanization) after China's economic reforms took off from the 1980s onwards accentuated the use of coal for industrialization, heating, and other purposes, contributing to regional pollution. This was the same case for Japan during its own rapid industrial growth in the 1960s (the "income-doubling" period).

The Case Study of South Korea

In the South Korean case, pollution in the country was at its most serious stage in the 1960s and 1970s due to accelerated economic growth and industrialization without due regard to environmental regulations, motivating the state to establish government departments specifically focused on managing the environment including the Environmental Agency in 1990.[10] At that time, the South Korean government also needed to persuade its households to switch to other fuels for domestic heating. In the mid-1990s, coal briquettes made up the biggest share of home heating

[8] Taek-Whan Han, "Northeast Asia Environmental Cooperation: Progress and Prospects" dated 23 September 1994 in the Nautilus Institute website [downloaded on 1 Jan 2017], available at http://nautilus.org/trade-and-environment/northeast-asia-environmental-cooperation-progress-and-prospects-3/

[9] Taek-Whan Han, "Northeast Asia Environmental Cooperation: Progress and Prospects" dated 23 September 1994 in the Nautilus Institute website [downloaded on 1 Jan 2017], available at http://nautilus.org/trade-and-environment/northeast-asia-environmental-cooperation-progress-and-prospects-3/

[10] Taek-Whan Han, "Northeast Asia Environmental Cooperation: Progress and Prospects" dated 23 September 1994 in the Nautilus Institute website [downloaded on 1 Jan 2017], available at http://nautilus.org/trade-and-environment/northeast-asia-environmental-cooperation-progress-and-prospects-3/

fuel in South Korea, meaning a disproportionate number of South Korean households used this cheap but polluting source for heating.[11] The population density of Northeast Asian cities (especially the first-tier cities in the region) also meant that pollution was concentrated and dense in certain areas. This has led to the rising middle classes in those cities to express socio-political dissatisfaction, compelling regional and national governments to react and come up with policies to reduce pollution or to enforce penalties for anti-pollution laws.

The Case Study of Japan

Japan's coal is regarded as high quality and, in 1991, when its domestic mines were closing, Japan's coal resources is approximated at 10 GT (gigatons) and recoverable reserves at below 1 GT.[12] But due to high manpower costs, depletion, and political strikes, Japan has stopped all large-scale exploitation of its domestic coal mines. Despite the shutdown of its own domestic coal mines, Japan's demand for coal energy is likely to remain relentless. Japan has in the pipeline 45 brand-new High-Energy, Low-Emissions (HELE) coal-fuelled electricity-generating plants that are likely to be around for the next three decades and uses high-grade black coal with Australia as one of its major suppliers of these high-quality coal, making up 33.3% of Japan's coal demand.[13] These plants are also described as ultra-supercritical coal-fired plants that can lessen emissions by 26% by the year 2030 and has stronger energy efficiency or emissions per megawatt (MW) hour by using technologies that combust coal at higher temperatures and mixture of gases in the combustion chamber.[14]

[11] Taek-Whan Han, "Northeast Asia Environmental Cooperation: Progress and Prospects" dated 23 September 1994 in the Nautilus Institute website [downloaded on 1 Jan 2017], available at http://nautilus.org/trade-and-environment/northeast-asia-environmental-cooperation-progress-and-prospects-3/

[12] Streets, David G., "Energy and Acid Rain Projections for Northeast Asia" dated 1 July 1991 in the International University of Japan GLOCOM website [downloaded on 1 Jan 2017], available at http://www.glocom.ac.jp/column/1991/07/energy_and_acid_rain_projectio.html

[13] McHugh, Babs, "Japanese government planning to build 45 new coal fired power stations to diversify supply" dated 1 February 2017 in the ABC website [downloaded on 1 Feb 2017], available at http://www.abc.net.au/news/rural/2017-01-31/japan-coal-power-plants/8224302

[14] Hewett, Jennifer, "Japan has good reason to burn coal for electricity" dated 3 Feb 2017 in The Australian Financial Review website [downloaded on 3 Feb 2017], available at http://www.afr.com/business/mining/coal/japan-has-good-reason-to-burn-coal-for-electricity-20170202-gu4csz

The Case Study of China

China's own growing need for coal will continue to drive its demand for domestic coal resources and even increase its imports. About 70% of China's energy demand is met by its own coal resources (approximately 114 billion tons in terms of total verified basic reserves of coal) and growing imports.[15] To facilitate this growing demand for coal, Dalian for example has become a growing centre of coal trading for China. Due to its own increasing demand, China is the world's largest producer of coal as well, from prolific mines found in the northern as well as northwestern regions of China. The "Made in China 2025 blueprint" inked by the State Council in May 2015 is keen to modernize/upgrade China's manufacturing sector through innovation and the use of Information Technology (IT) in the manufacturing sector.[16]

Simultaneously, China will facilitate efficiency increases and getting rid of wastes and obsolete heavy manufacturing technologies/activities, focusing coal utilization in big-scale efficient power and heat generation facilities while mothballing hundreds of thousands of inefficient, small-scale polluting coal-fired boilers.[17] The "Energy Development Strategy Action Plan (2014–2020)" published by the State Council prescribes restrictions on annual primary energy and coal consumption until 2020 at absolute levels and sets quotas for lessening the coal share of primary energy consumption to less than 62% and increasing the percentage of use of non-fossil energy to 15% by 2020 and to about 20% by 2030.[18] Therefore, the coal percentage of China's electricity generation is

[15] Campi, Alicia, "The New Great Game in Northeast Asia: Potential Impact of Energy Mineral Development in Mongolia on China, Russia, Japan, and Korea" dated 2013 in the Asia-Pacific Policy Papers Series [downloaded on 1 Jan 2017], (Washington, DC: Edwin O. Reischauer Center for East Asian Studies), 2013, available at http://www.reischauer-center.org/en/wp-content/uploads/2013/06/RC-Monograph-2013-Campi_The-New-Great-Game-in-Northeast-Asia.pdf, p. 12.

[16] US Energy Information Administration, "Chapter 4. Coal" dated 11 May 2016 in International Energy Outlook 2016 [downloaded on 1 Jan 2017], available at https://www.eia.gov/outlooks/ieo/coal.php

[17] US Energy Information Administration, "Chapter 4. Coal" dated 11 May 2016 in International Energy Outlook 2016 [downloaded on 1 Jan 2017], available at https://www.eia.gov/outlooks/ieo/coal.php

[18] US Energy Information Administration, "Chapter 4. Coal" dated 11 May 2016 in International Energy Outlook 2016 [downloaded on 1 Jan 2017], available at https://www.eia.gov/outlooks/ieo/coal.php

expected to go down from about 75% in 2012 to 45% in 2040, and the coal share of China's total primary energy consumption from 66% in 2012 to 44% in 2040.[19]

Coal and Its Importance in China's Energy Mix

The People's Republic of China greatly shapes global coal prices as well as supply and demand through its immense use of the resource to produce coke oven coke, crude steel, pig iron, and cement. To mitigate pollution, China is following a regional trend of shutting down small-scale coal-powered generators in favour of larger and more efficient power generation plants. Besides corruption issues, infrastructure evaluations, and political campaigns in the Chinese coal industry, the socio-economic effects and environmental impacts of coal use have also become significant topic in political discussions. When Chinese coal supply was reduced in the fall of 2016 through various means such as the restricting of trucking venues, prices of coal went up, but when the restrictions have been lifted to cope with demand, the increase in prices causes coal prices to drop.

There is hence a need to mitigate price fluctuations. China's energy demand is met by its own coal resources and growing imports. Three provinces stand out in domestic coal supply: Shanxi, Shaanxi, and Inner Mongolia. The "Made in China 2025 blueprint" inked by the State Council in May 2015 aims to modernize/upgrade China's manufacturing sector through focusing on coal utilization in big-scale power and heat generation facilities. The "Energy Development Strategy Action Plan (2014–2020)" published by the State Council prescribes restrictions on annual primary energy and coal consumption until 2020. Measures designed to cope with excess overcapacity in steel and coal production have generated images and concerns of a declining coal belt, raising concerns about the local economies of the affected coal mining towns.

Coal however remains responsible for approximately 75% of China's electricity power in spite of large hydroelectric dam constructions and initiatives to implement solar panels and wind turbines. Chinese coal investments overseas have increased to procure more resources to generate greater electricity volume. The United Nations (UN) lists China as the

[19] US Energy Information Administration, "Chapter 4. Coal" dated 11 May 2016 in International Energy Outlook 2016 [downloaded on 1 Jan 2017], available at https://www.eia.gov/outlooks/ieo/coal.php

world's biggest investor in renewable energy projects since at least 2011. North Korea is an important supplier of coking coal for steelmaking in China. The cheap North Korean coal however comes with political repercussions for China from the international community. China is also making coal investments in Southeast Asia to meet its coal needs, an issue discussed later in the chapter.

Recent Developments in China's Coal Use

The single most important trend in the history of the global coal industry is the rise of demand from the People's Republic of China (PRC). China's use of coal has seen a spike of 392% from the year 2000 to hit 611.1 MT (megatons) in 2015 and in the process expanded its share in global output from 26% (2000) to 56.1% (2015).[20] China greatly shapes global coal prices as well as supply and demand through its immense use of the resource. It uses coal to make 468 MT of coke oven coke (66.0% of global volume), 823 MT of crude steel (49.3% of international output), 712 MT of pig iron (60.1% of global production), and approximately 2.50 GT of cement (59.8% share of the world's production).[21] However, massive use comes with its own set of challenges. Coal pollutes compared to other fossil fuels like oil and natural gas. To mitigate pollution, China is following a regional trend of shutting down small-scale coal-powered generators in favour of larger and more efficient power generation plants.

China's growing need for coal will continue to drive its demand for domestic coal resources and even increase its imports. Seventy per cent of China's energy demand is met by its own coal resources (approximately 114 billion tons in terms of total verified basic reserves of coal) and growing imports.[22] China has continued to be the globe's largest coal producer since 1985 with a total production of 3.527.2 MT of coal.[23] Currently

[20] International Energy Agency (IEA), Key Coal Trends Excerpt from: Coal Information 2016 (Paris: IEA), 2016, p. 5.

[21] International Energy Agency (IEA), Key Coal Trends Excerpt from: Coal Information 2016 (Paris: IEA), 2016, p. 9.

[22] Campi, Alicia, "The New Great Game in Northeast Asia: Potential Impact of Energy Mineral Development in Mongolia on China, Russia, Japan, and Korea" dated 2013 in the Asia-Pacific Policy Papers Series [downloaded on 1 January 2017], (Washington, DC: Edwin O. Reischauer Centre for East Asian Studies), 2013, available at http://www.reischauer-center.org/en/wp-content/uploads/2013/06/RC-Monograph-2013-Campi_The-New-Great-Game-in-Northeast-Asia.pdf, p. 12.

[23] International Energy Agency (IEA), Key Coal Trends Excerpt from: Coal Information 2016 (Paris: IEA), 2016, p. 3.

Qinhuangdao is the most important port for China's coal trade and the annual coal trade fair organized in the city in November 2016 was a facility for coal producers and electricity power plant managers to conclude more long-term deals overseen by China National Coal Association (CNCA).[24] From 1949, Qinhuangdao's major ice-free harbour facilities have been optimally utilized, exporting coal, coke, petroleum, and timber and, in the case of coal, the port handles a large share of the overall national coal petroleum freightage.[25] Qinhuangdao, as China's main coal port, saw China's coal imports increased 33% year on year (y/y) to 89.5 million tons in the first quarter of 2017 due to electricity power generation.[26]

Dalian for example has become a growing centre of coal trading for China. Due to its own increasing demand, China is the world's largest producer of coal as well, from prolific mines found in the northern as well as northwestern region of China. Dalian was the location of the Fourth Northeast Asia Coal Trade Fair on 1 August 2013. Managers of approximately 600 firms that manufacture or distribute coal were hosted by CNCA, China Coal Transportation and Sale Society, and the Northeast Asia Coal Trade Centre.[27] Dalian has also become a major coal futures trading platform. The Dalian Commodity Exchange is now a premium platform for coking coal futures with more than one-million contracts of iron ore futures transacting on a daily basis.[28]

The directives of the "Made in China 2025 blueprint" inked by the State Council in May 2015 to modernize/upgrade China's manufacturing sector through innovation and the use of Information Technology (IT) in

[24] *Global Times*, "Annual coal trade fair in Qinhuangdao" dated 27 November 2016 in the Global Times website [downloaded on 1 January 2017], available at http://www.globaltimes.cn/content/1020536.shtmlQinhuangdao

[25] The Editors of Encyclopædia Britannica, "Qinhuangdao (Alternative Title: Ch'in-huang-tao)" in Encyclopaedia Britannica [downloaded on 1 August 2017], available at https://www.britannica.com/place/Qinhuangdao China' s coal imports up 33% to nearly 90 million tonnes in Q1.

[26] Zeng, Xiaolin, "Coal mining in China" dated 30 May 2017 in the IHS Fairplay: Maritime Shipping News website [downloaded on 30 May 2017], available at https://fairplay.ihs.com/commerce/article/4287086/china-s-coal-imports-up-33-to-nearly-90-million-tonnes-in-q1

[27] Zhao, Qian and Roger Bradshaw (editors), "Dalian plays host to Northeast Asia Coal Trade Fair" dated 6 August 2013 in China Daily [downloaded on 1 January 2017], available at http://www.chinadaily.com.cn/m/dalian/2013-08/06/content_16874898.htmEXCHANGES

[28] Tan, Huileng, "China's wild futures trading is opening up a major opportunity for exchanges abroad" dated 21 May 2017 in the CNBC.com website [downloaded on 21 May 2017], available at https://www.cnbc.com/2017/05/21/heres-how-commodity-exchanges-are-eyeing-chinas-wild-futures-trade.htm

the manufacturing sector may impact the coal industry.[29] China will facilitate efficiency increases and get rid of wastes and obsolete heavy manufacturing technologies/activities, focusing coal utilization in big-scale efficient power and heat generation facilities while mothballing hundreds of thousands of inefficient, small-scale polluting coal-fired boilers.[30] The "Energy Development Strategy Action Plan (2014–2020)" published by the State Council prescribes restrictions on annual primary energy and coal consumption until 2020 at absolute levels. It also sets quotas for lessening the coal share of primary energy consumption to less than 62% and for increasing the percentage of the use of non-fossil energy to 15% by 2020 and to about 20% by 2030.[31] Therefore, the coal percentage of China's electricity generation is expected to go down from about 75% in 2012 to 45% in 2040 and likewise for the coal share of China's total primary energy consumption from 66% in 2012 to 44% in 2040.[32] Coal's percentage of total energy consumption dropped from 64.0% in 2015 to 62.0% in 2016.[33]

The Politics of Coal in China

Like most other resource industries in China, the coal industry has its own set of political considerations. Technocrats face the scrutiny of the party political leadership and may be affected by involvement in corruption cases or perceptions of not doing enough to solve coal industry problems. Under Chinese President Xi Jinping's anti-corruption campaign, in 2015,

[29] US Energy Information Administration, "Chapter 4. Coal" dated 11 May 2016 in *International Energy Outlook 2016* [downloaded on 1 January 2017], available at https://www.eia.gov/outlooks/ieo/coal.php

[30] US Energy Information Administration, "Chapter 4. Coal" dated 11 May 2016 in International Energy Outlook 2016 [downloaded on 1 January 2017], available at https://www.eia.gov/outlooks/ieo/coal.php

[31] US Energy Information Administration, "Chapter 4. Coal" dated 11 May 2016 in International Energy Outlook 2016 [downloaded on 1 January 2017], available at https://www.eia.gov/outlooks/ieo/coal.php

[32] US Energy Information Administration, "Chapter 4. Coal" dated 11 May 2016 in International Energy Outlook 2016 [downloaded on 1 January 2017], available at https://www.eia.gov/outlooks/ieo/coal.php

[33] Lin, Alvin, "Understanding China's New Mandatory 58% Coal Cap Target" dated 17 March 2017 in Natural Resources Defence Council (NRDC) website [downloaded on 17 March 2017], available at https://www.nrdc.org/experts/alvin-lin/understanding-chinas-new-mandatory-58-coal-cap-target

the Baoding City Intermediate People's Court sentenced Wei Pengyuan, former deputy chief of the coal bureau under the National Energy Administration, for accepting 211 million yuan ($32.5 million) in bribes and failing to account for 131 million yuan ($20.2 million) worth of missing assets.[34] Even the development of coal mining cities is subjected to political scrutiny. For example, critics of Datong's excess infrastructure building claimed that the city's former Mayor Geng Yanbo (nicknamed "Demolition Geng," "Geng ChaiChai," or "Geng Smash-Smash") from 2008 to early 2013 was responsible for overbuilding of infrastructure including a 3.3 square km of "ancient city" replica that relocated 500,000 residents to make way for the project.[35] Geng's legacy has both supporters and detractors. Geng's critics like Renmin University economics Professor Tao Ran called his programme excessive, but supporters of Geng considered him a visionary mayor and Director Zhou Hao even shot a Golden Horse-winning documentary titled *The Chinese Mayor* based on Geng who was eventually promoted in 2013 to Taiyuan mayor in Shanxi's provincial capital.[36]

Besides corruption issues, infrastructure evaluations, and political campaigns, socio-economic effects and environmental impacts of coal use have also become political fodder. Environmental political pressure on the coal industry also comes from the highest levels of government as Premier Li Keqiang publicly announced the leadership's intentions to "make our skies blue again."[37] On the other hand, coal stakeholders warned of the social impact of closing down coal mines. In the National People's Congress (NPC) session in March 2016, Datong Coal Chairperson Zhang

[34] Associated Press, "Chinese Coal Official Who Hid $30 M Admits to Corruption" dated 29 December 2015 in the VOA website [downloaded on 1 January 2017], available at https://www.voanews.com/a/china-corruption/3122856.html

[35] Zhou, Xin, "Decline and fall: the broken dreams of a Chinese coal-mining city struggling to address industrial overcapacity" dated 30 May 2017 in the *South China Morning Post* [downloaded on 30 May 2017], available at http://www.scmp.com/news/china/economy/article/1935326/decline-and-fall-broken-dreams-chinese-coal-mining-city

[36] Zhou, Xin, "Decline and fall: the broken dreams of a Chinese coal-mining city struggling to address industrial overcapacity" dated 30 May 2017 in the *South China Morning Post* [downloaded on 30 May 2017], available at http://www.scmp.com/news/china/economy/article/1935326/decline-and-fall-broken-dreams-chinese-coal-mining-city

[37] Bradshernov, Keith, "Despite Climate Change Vow, China Pushes to Dig More Coal" dated 29 November 2016 in the *New York Times* [downloaded on 1 January 2017], available at https://www.nytimes.com/2016/11/29/business/energy-environment/china-coal-climate-change.html?mcubz=1

Youxi urged the government to look into the impact of projected large-scale retrenchments, citing how coal mines tend to hire 5000–7000 employees with approximately 20,000 dependent relatives, numerically equating mine closures with closures of cities and towns.[38]

In response to the top leadership's call and popular demands from an increasingly vocal middle class in 2016 for cleaner air and to deal with overcapacity, the National Development and Reform Commission (NDRC) further ceased the building of 50 Gigawatts (GW) of coal power plants in 2016 and, in 2015, China closed down 290 million tons of out-dated coal capacity.[39] In 2016, the Ministry of Environmental Protection stopped the use of trucks to transport coal at the port of Tianjin and closed the trucking route from Hebei's coal railway to ships in September 2016 in a bid to reduce coal supply and use in China. Sceptics however argue that legislations can be diluted by ministries and industries at the local and provincial levels.[40]

Cutting down or increasing coal supply also comes with its own set of political economic considerations. When coal supply was reduced in the fall of 2016 through various means like restricting trucking venues, prices of coal went up. However, when the restrictions are lifted to cope with demand, the increase in prices causes coal price drop. In early 2017, drop in coal prices prompted senior management from 19 largest coal firms in China (including Shenhua) to lobby the government to reintroduced controls on coal output like the case in late 2016 after the peak demands of winter were over in China.[41] These 19 de facto coal industry sector lobbyists are also dialoguing with the state to reintroduce the limits on mines' output, allowing mines to

[38] Zhou, Xin, "Decline and fall: the broken dreams of a Chinese coal-mining city struggling to address industrial overcapacity" dated 30 May 2017 in the *South China Morning Post* [downloaded on 30 May 2017], available at http://www.scmp.com/news/china/economy/article/1935326/decline-and-fall-broken-dreams-chinese-coal-mining-city

[39] Meidan, Michael, "Political Transition Will Override China's Policy Targets" dated 17 March 2017 in the Chatham House website [downloaded on 17 March 2017], available at https://www.chathamhouse.org/expert/comment/political-transition-will-override-china-s-policy-targets

[40] Meidan, Michael, "Political Transition Will Override China's Policy Targets" dated 17 March 2017 in the Chatham House website [downloaded on 17 March 2017], available at https://www.chathamhouse.org/expert/comment/political-transition-will-override-china-s-policy-targets

[41] Tabeta Shunsuke, "China moving to renew coal mining curbs" dated 23 February 2017 in Nikkei Asia [downloaded on 23 February 2017], available at https://asia.nikkei.com/Politics-Economy/Economy/China-moving-to-renew-coal-mining-curbs

operate only for a maximum of 276 days annually.[42] The coal mining sector (mostly state-owned enterprises, SOEs) in China fears an oversupply of coal will affect prices if no cuts are implemented; the state's priority on the other hand is price stability to ward off any potential social disturbance or dissatisfaction that can arise from inflation.[43]

Is the Coal Belt Declining?

The Chinese state is reducing coal's contribution to its energy mix from 64.5% (2015) to 50% by 2030 to upgrade air quality and to cope with an overcapacity in coal and steel production as the need for electric power using coal fuels is also slowing down (coal prices have dropped by more than 50% between 2011 and 2016), alongside a maturing economy.[44] These measures designed to cope with excess overcapacity in steel and coal production have generated images and concerns of a declining coal belt, raising concerns about the impact it will have on the local economies of the affected coal mining towns. Datong, a 1.7 million strong city located 350 km west of Beijing and the base of the massive state-owned Datong Coal Mine Group with 175,000 workers, is representative of a stereotypical coal rust belt in China combating excess industrial capacity.[45]

The city is facing impending large-scale retrenchments, unpaid salaries for miners, uncompleted infrastructures due to sagging demand, an underused multimillion-dollar sports stadium, high business taxes, property glut, risky financial situation, overreliance on coal, and protests by thousands of coal mining workers in Shuangyashan bordering Russia.[46]

[42] Tabeta Shunsuke, "China moving to renew coal mining curbs" dated 23 February 2017 in Nikkei Asia [downloaded on 23 February 2017], available at https://asia.nikkei.com/Politics-Economy/Economy/China-moving-to-renew-coal-mining-curbs

[43] Tabeta Shunsuke, "China moving to renew coal mining curbs" dated 23 February 2017 in Nikkei Asia [downloaded on 23 Feb 2017], available at https://asia.nikkei.com/Politics-Economy/Economy/China-moving-to-renew-coal-mining-curbs

[44] Zhou, Xin, "Decline and fall: the broken dreams of a Chinese coal-mining city struggling to address industrial overcapacity" dated 30 May 2017 in the *South China Morning Post* [downloaded on 30 May 2017], available at http://www.scmp.com/news/china/economy/article/1935326/decline-and-fall-broken-dreams-chinese-coal-mining-city

[45] Zhou, Xin, "Decline and fall: the broken dreams of a Chinese coal-mining city struggling to address industrial overcapacity" dated 30 May 2017 in the *South China Morning Post* [downloaded on 30 May 2017], available at http://www.scmp.com/news/china/economy/article/1935326/decline-and-fall-broken-dreams-chinese-coal-mining-city

[46] Zhou, Xin, "Decline and fall: the broken dreams of a Chinese coal-mining city struggling to address industrial overcapacity" dated 30 May 2017 in the *South China Morning*

Datong's Yudong New District has witnessed abandoned high-rise buildings, an almost-finished 1.2 billion yuan sports stadium exposed to weather conditions, deserted library, opera house and fine-art museum, and an inactive China Tang City; Datong's overall property investment declined by 41.4% in 2015 as the coal industry faced smaller demand.[47] However, to say that the coal industry faces an irreversible and absolute decline is not accurate as coal fuels will continue to be an important part of China's overall energy mix.

Coal Industry in China: Still Critical Despite Its Tremendous Advance in Hydro and Solar Energy?

Coal remains responsible for approximately 75% of China's electricity power as of late 2016 in spite of large hydroelectric dam constructions and the globe's biggest initiative to implement solar panels and construct wind turbines infrastructure.[48] China's National Action Plan for Air Pollution Prevention and Control's (NAPAPPC) mid-term review published on 5 July 2016 indicated eight major regions increased 50.8 GW of new coal-fired energy capacity from 2013 to 2015; in 2016, 42 GW of coal-fired capacity was work in progress and 11 GW of power was cleared for construction in 2015.[49]

By comparison, from 2013 to 2015, only 10.8 GW of coal-fired capacity in eight provinces was reduced. Simply put, the increase in demand for coal-generated power outstripped the reduction in the same

Post [downloaded on 30 May 2017], available at http://www.scmp.com/news/china/economy/article/1935326/decline-and-fall-broken-dreams-chinese-coal-mining-city

[47] Zhou, Xin, "Decline and fall: the broken dreams of a Chinese coal-mining city struggling to address industrial overcapacity" dated 30 May 2017 in the *South China Morning Post* [downloaded on 30 May 2017], available at http://www.scmp.com/news/china/economy/article/1935326/decline-and-fall-broken-dreams-chinese-coal-mining-city

[48] Bradshernov, Keith, "Despite Climate Change Vow, China Pushes to Dig More Coal" dated 29 Nov 2016 in the New York Times [downloaded on 1 Jan 2017], available at https://www.nytimes.com/2016/11/29/business/energy-environment/china-coal-climate-change.html?mcubz=1

[49] Shepard, Wade, "If China Is So Committed To Renewable Energy, Why Are So Many New Coal Plants Being Built?" dated 8 July 2016 in Forbes.com [downloaded on 8 July 2016], available at https://www.forbes.com/sites/wadeshepard/2016/07/08/if-china-is-so-committed-to-renewable-energy-why-are-so-many-new-coal-plants-being-built/#36dbe1235918

category, indicating the continued importance of coal as an energy resource. Given that these investments are made by China, coal will continue to be a fuel of choice for Chinese power plants as coal power plants are long-term investments and each power plant can operate for 30–50 years.[50] China's continual coal power plant investment is to feed its increasing energy consumption. Dwarfing both Japan and South Korea, China is the biggest user of coal globally, consuming 76 quadrillion Btu of coal in 2012, making up 50% of global coal consumption. Current trends are showing signs of slowing as China adjusts to the "new normal" in economic development and as the energy industry goes through fundamental transformation to deal with politically sensitive environmental problems.[51]

The slowdown is visible in the most recent decline of 0.19 billion tons (physical tons) from 2015 to 2016.[52] Part of the slowdown is due to the impact of China switching to renewable energy fuels. By 2020, more than 15% of China's energy is projected to come from non-fossil fuels such as wind, solar, and hydropower, while, ironically at the same time, it will also be the largest coal user in the world.[53] In the long-term scenario, however, China is investing and switching to renewable energy, moving from polluting affordable coal to high-grade ultra-supercritical coal that can combust at a very hot temperature with a high level of efficiency, using some of the newest equipment in the world.

[50] Shepard, Wade, "If China Is So Committed To Renewable Energy, Why Are So Many New Coal Plants Being Built?" dated 8 July 2016 in Forbes.com [downloaded on 8 July 2016], available at https://www.forbes.com/sites/wadeshepard/2016/07/08/if-china-is-so-committed-to-renewable-energy-why-are-so-many-new-coal-plants-being-built/#36dbe1235918

[51] US Energy Information Administration, "Chapter 4. Coal" dated 11 May 2016 in International Energy Outlook 2016 [downloaded on 1 January 2017], available at https://www.eia.gov/outlooks/ieo/coal.php

[52] The author calculated this figure based on the data available at Lin, Alvin, "Understanding China's New Mandatory 58% Coal Cap Target" dated 17 March 2017 in Natural Resources Defence Council (NRDC) website [downloaded on 17 March 2017], available at https://www.nrdc.org/experts/alvin-lin/understanding-chinas-new-mandatory-58-coal-cap-target

[53] Shepard, Wade, "If China Is So Committed To Renewable Energy, Why Are So Many New Coal Plants Being Built?" dated 8 July 2016 in Forbes.com [downloaded on 8 July 2016], available at https://www.forbes.com/sites/wadeshepard/2016/07/08/if-china-is-so-committed-to-renewable-energy-why-are-so-many-new-coal-plants-being-built/#36dbe1235918

China as a Regional Coal Industry Investor and Importer

The United Nations (UN) lists China as the world's biggest investor in renewable energy projects since at least 2011,[54] albeit with coal remaining critical despite developments in renewable energies like hydropower and solar energy. Other measures to combat pollution issues have been adopted like conserving energy or efficient energy use to reduce wastage and lower energy consumption of energy and reduce pollution as well; substitution of a much cleaner fossil fuel of natural gas and/or renewable energy resources like hydropower, wind, solar, and geothermal; and turning to multiple sources of fossil fuels, including Australia, the Southeast Asian region, and North Korea in the neighbouring region. Southeast Asia is also potentially a destination for Chinese coal investments as well.

Coal Trade and Politics Between China and North Korea

Interestingly, North Korea has quite big reserves of coal approximated at 12 GT and numerous potentially undiscovered coal mines.[55] Its problems in coal extraction are manmade. Years of isolation, sanction, and self-reliance *juche* have not brought about rapid economic development or cutting-edge industrialization. North Korea is an important supplier of coking coal for making steel to China.[56] Its low-cost labour potentially makes North Korean coal cheap, but the payback for taking Pyongyang's resources is political repercussions. When the North Koreans tested a missile in early 2017, China turned back North Korean ships carrying coal for export. North Korea is under increasing international sanctions, even

[54] VOA, "Despite Renewable Energy Push, China Still Runs on Coal" dated 21 November 2011 in the VOA website [downloaded on 1 August 2017], available at https://www.voanews.com/a/despite-renewable-energy-push-china-still-runs-on-coal-134331173/168226.html

[55] Streets, David G., "Northeast Asia coal trading center set up in Dalian" dated 1 July 1991 in the GLOCOM website [downloaded on 1 Jan 2017], available at http://www.glocom.ac.jp/column/1991/07/energy_and_acid_rain_projectio.html, unpaginated.

[56] Reuters staff, "China says enforcing North Korea coal ban seriously, no violation" dated 21 April 2017 in Reuters [downloaded on 21 April 2017], available at http://www.reuters.com/article/us-storm-harvey/trump-to-visit-victims-of-unprecedented-floods-in-texas-and-louisiana-idUSKCN1BD074

from its ally China, for its missile testing and nuclear programmes. Its recent test fire of Intercontinental Ballistic Missiles (ICBMs) and threats to attack Guam did not go down well with the international community. The US has promised tougher crackdown on Pyongyang's trade lifeline and financial activities while opening the door for talks. These sanctions and tougher stance against Pyongyang is likely to stifle North Korean industries (including coal mining) even further.

There is an international dimension to the Sino-North Korean coal trade. On 5 August 2017 after the North Korean ICBM tests, the United Nations Security Council (UNSC) levelled new sanctions on Pyongyang with the support of China, but China also informed the US not to engage in a trade war with China on other fronts. It is unclear if this was a Chinese bargain of going hard on North Korea to exact concessions in its own trading relationship with the US. An important element in the international pressure on North Korea centres on coal trade and China has moved to stop the coal trade with North Korea even before the August 2017 UN sanctions. China makes up approximately 90% of North Korean trade and coal occupies 50% of overall North Korean exports and China had already suspended North Korean coal imports from February 2017 till the end of 2017.[57]

North Korea exports to China dropped to US $880 million between January and June 2017, which is a decline of 13% compared to the same months in 2016 and China's coal imports from North Korea declined sharply with a trickle of 2.7 million tons entering China in January–June 2017, a drop of 75% compared with 2016.[58] Enforcement of Chinese coal ban is closely scrutinized by the international community. Reuters spotted a few fully loaded North Korean cargo vessels returning to North Korea on 11 April 2017 after the Chinese customs imposed the obligations under the UN sanctions on 7 April 2017 and informed Chinese trading

[57] Feng, Shirley, Yang Liu and Luna Lin, "China bans North Korean iron, lead, coal imports as part of U.N. sanctions" dated 14 August 2017 in The Washington Post [downloaded from https://www.washingtonpost.com/world/china-bans-north-korea-iron-lead-coal-imports-as-part-of-un-sanctions/2017/08/14/a0ce4cb0-80ca-11e7-82a4-920da1aeb507_story.html?utm_term=.2045009a70ea]

[58] Feng, Shirley, Yang Liu and Luna Lin, "China bans North Korean iron, lead, coal imports as part of U.N. sanctions" dated 14 August 2017 in The Washington Post [downloaded from https://www.washingtonpost.com/world/china-bans-north-korea-iron-lead-coal-imports-as-part-of-un-sanctions/2017/08/14/a0ce4cb0-80ca-11e7-82a4-920da1aeb507_story.html?utm_term=.2045009a70ea]

companies to turn away North Korean coal shipments.[59] But, on 21 April 2017, NKNews.org released news (which was picked up by the international media) that a few North Korean vessels were spotted in the vicinity of Tangshan northern Chinese port. Questions about such sightings triggered off Chinese Foreign Ministry spokesperson Lu Kang's response that China was "seriously enforcing" the ban on North Korean coal imports and that it may be possible that some North Korean mariners may "need to be looked after for humanitarian reasons," accounting for the presence of those ships.[60]

China's Coal Investments in Southeast Asia

A promising region for coal trade and business is Southeast Asia. One of the primary targets for such investments in Southeast Asia is Indonesia. According to the *Time* article "King Coal's Comeback" dated 11 June 2012, China currently imports most of its foreign coal from Indonesia and Australia.[61] Due to supply potential of the Southeast Asian region, Chinese state-owned enterprises (SOEs) and government loans teamed up to invest in Southeast Asian economies. While China is keen to shore up its credentials as the world's environmental power after the US backed off the Paris climate agreement and is actively adopting cleaner energy sources like renewables to lessen dependence on coal, its coal investments overseas have increased. At the point of this writing, Chinese financial institutions and enterprises are involved in a minimum of 79 coal electricity generation facility/projects adding up to 52 GW or more in electricity volume.[62] Such investment plans fit into Southeast Asian energy priorities

[59] Reuters staff, "China says enforcing North Korea coal ban seriously, no violation" dated 21 April 2017 in Reuters [downloaded on 21 April 2017], available at http://www.reuters.com/article/us-storm-harvey/trump-to-visit-victims-of-unprecedented-floods-in-texas-and-louisiana-idUSKCN1BD074

[60] Reuters staff, "China says enforcing North Korea coal ban seriously, no violation" dated 21 April 2017 in Reuters [downloaded on 21 April 2017], available at http://www.reuters.com/article/us-storm-harvey/trump-to-visit-victims-of-unprecedented-floods-in-texas-and-louisiana-idUSKCN1BD074

[61] Walsh, Bryan, "King Coal's Comeback" dated 11 June 2012 in Time (NY: Time Asia), 2012, p. 41.

[62] Walker, Beth, "Chinese Investment Stokes Global Coal Growth" dated 24 September 2016 in The Diplomat [downloaded on 1 Jan 2017], available at http://thediplomat.com/2016/09/chinese-investment-stokes-global-coal-growth/

because electricity needs are poised to go up by 83% from 2011 to 2035 and a large number of Southeast Asian countries are still using coal-powered electricity generation facilities.[63]

The Case of Indonesia China appears to find coal energy complementarity and synergy with Indonesia, building power plants in the world's largest Muslim country as the world's most populated country also buys coal resources from Indonesia. Indonesia is the cornerstone of Chinese President Xi Jinping's One Belt, One Road initiative (OBOR, now known as Belt and Road Initiative or BRI). It is an archipelago with vast resources spread over 13,000 islands. There seems to be some matching demand and supply of coal and coal-powered electricity generation between the two countries. The Chinese viewpoint detects economic opportunities in the form of electricity shortages in Indonesia, and China is keen to supply its technologies to help Indonesia cope better with the situation. From the Chinese perspective, it appears electricity shortage is holding back Indonesian economic development. There is also a dualistic approach in carrying out their coal investments through state-owned enterprises (SOEs) and private initiatives. The SOE route is necessary given the scale of the engagement, building some of the largest power plants in Indonesia. An advantage in working at the government-to-government (G2G) project state level is the economy of scale, with the Chinese forgoing any forms of additional incentives from the Indonesian government for building the mega projects.[64] Private sector initiatives are smaller and nimbler and can add a private sector contour to the Belt and Road Initiative (BRI). It may also be more market sensitive and can harness profitability as a priority in doing business.

From the Indonesian perspective, job creation, consistent access to electricity, energy provision spurring economic development and industrialization, and the ability to power mineral extraction are major priorities for the government. Job creation also has an additional spin-off in terms of generating tax revenue for the government, including the coffers of

[63] Myllyvirta, Lauri, "Southeast Asia is planning 400 new coal power plants—what does that mean?" dated 13 January 2017 in the Energydesk Greenpeace website [downloaded on 13 Jan 2017], available at http://energydesk.greenpeace.org/2017/01/13/southeast-asia-coal-plans-health-japan-indonesia/

[64] Jakarta Globe, "China Invests $17b in Indonesian Power Project" dated 29 May 2013 in the Jakarta Globe website [downloaded on 1 Jan 2018], available at http://jakartaglobe.id/business/china-invests-17b-in-power-project/

provincial governments in Indonesia. There is an ambition to bring Indonesia into the middle-income country status and become a major economy in the world. Unlike Myanmar which is connected continentally to China, Indonesia is a disparate far-flung archipelago, so the prospects of China building power plants in Indonesia to supply energy and power to China is comparatively more remote. Thus, Chinese importation of energy from Indonesia is still mainly in the form of fossil fuels. But, as Indonesia's economy takes off, it also requires energy resources for its own economic development. Thus, China's role in supply upstream and sophisticated coal-fired power plants will become increasingly important, more important than just buying coal commodities from Indonesia.

In other words, the relationship will slowly morph from coal consumer and supplier relationship to technological transfers using the turnkey format and in a way complementary to both parties. The bilateral relationship in this sense is becoming more value-added and will gradually shift from fossil fuel (especially coal)-dependent to a mutually reinforcing one that will capitalize on the core strengths of both economies. There are three phases involved in this process. The first phase is negotiations and talks for Chinese state-owned enterprises to propose and sell coal power plants to the Indonesians. This phase took place before 2015. The second stage consists of the period 2016–2020 when some of these Chinese-built coal-powered projects become operational. And the third phase is from 2019 to 2020 and beyond when all the envisioned plants will be progressively completed in phased format. By then, the bilateral relationship may once again alter and trend towards more value-added exchanges.

In 2013, Vice President Xia Zhong of Chinese state-owned enterprise (SOE) China Power Investment Corporation and Indonesian Energy and Mineral Resources Minister Jero Wacik outlined US $17 billion investments in Indonesia to build its biggest power plants (including the 7000-MW hydropower plants in Kalimantan slated for completion in 2020 that will generate power from the Tayan River in North Kalimantan).[65] China Power's main products and services are centred in power generation and have a total electric power capacity of 80,074 MW (including a 6000 MW

[65] Jakarta Globe, "China Invests $17b in Indonesian Power Project" dated 29 May 2013 in the Jakarta Globe website [downloaded on 1 Jan 2018], available at http://jakartaglobe.id/business/china-invests-17b-in-power-project/

hydropower plant in Myanmar for exporting electricity to China).[66] In 2016, work started on the first China-built coal power plant in Bengkulu, the 2 × 100-MW US $360 million Tenaga Listrik Bengkulu (TLB), a joint venture (JV) between Indonesian publicly listed firm PT. Intraco Penta and the Chinese state-owned enterprise (SOE) Power Construction Corporation of China (PowerChina that won the public bid in November 2015) slated for completion in 2019.[67] The deal involved a 25-year duration turnkey build-own-operate-transfer (BOOT) format that fits into Indonesian President Joko Widodo's goal of increasing 35,000 MW of power from 2016 to 2021.

The Indonesian local authorities echo Chinese sentiments of electricity shortage holding back economic development as Bengkulu Governor Ridwan Mukti said that though Bengkulu has plentiful resources like coal, fishery, and agriculture, power supply shortage has held back investments and fomented high poverty rate.[68] In this way, China's own developmental experiences had some parallels. China decided to embark on economic reforms in the late 1970s and discovered the importance of basic infrastructure as a necessary ingredient for economic development. Such parallel experiences were revealed in statements made by Chinese private sector leaders:

> According to Zhan Ke, president director of PT. GCL Indo Tenaga, a joint venture between GCL and Indonesia Power, the population of Indonesia is three times that of eastern China's Jiangsu province, yet the installed power-generation capacity is only 560,000 megawatts, half of that in Jiangsu province.[69]

[66] Jakarta Globe, "China Invests $17b in Indonesian Power Project" dated 29 May 2013 in the Jakarta Globe website [downloaded on 1 Jan 2018], available at http://jakartaglobe.id/business/china-invests-17b-in-power-project/

[67] Xinhua, "China-invested power plant starts construction in Indonesia to resolve electricity shortage" dated 25 Oct 2016 in the Xinhuanet website [downloaded on 25 Oct 2016], available at http://www.xinhuanet.com/english/2016-10/25/c_135780336.htm

[68] Xinhua, "China-invested power plant starts construction in Indonesia to resolve electricity shortage" dated 25 Oct 2016 in the Xinhuanet website [downloaded on 25 Oct 2016], available at http://www.xinhuanet.com/english/2016-10/25/c_135780336.htm

[69] Zheng, Xin and Zou Shuo, "Coal-fired power project kicks off" dated 10 August 2017 in China Daily [downloaded on 10 August 2017], available at http://www.chinadaily.com.cn/business/2017-08/10/content_30402968.htm

Indonesian power utility authority Perusahaan Listrik Negara (PLN) revealed the project in western Sumatra can provide approximately 1.4 Gigawatt Hour (GWh) of electricity to Sumatra power grid and shave off 1.7 trillion rupiah (approximately US $130 million) annually after it is up and running.[70] About 75% of the funding for the project came from the Industrial and Commercial Bank of China (ICBC) and Export-Import Bank of China and the remainder is from private equity of the project stakeholders.[71]

In the same year in 2016, another project 200 MW (1.4 GW hourly) Kalbar-1 power station managed by Suzhou commercial enterprise Golden Concord Holdings Ltd. (GCL) working with Indonesia Power (under the Indonesian power utility group Perusahaan Listrik Negara) started in West Kalimantan province is slated for completion in 2020.[72] This project is unique because it is a private sector initiative rather than an SOE-led consortium; in fact, GCL is the inaugural commercial firm from China to ink a power-purchase deal with Indonesia. This unique case study is a valuable one as the company is inserting this project under the rubric of Belt and Road Initiative (BRI):

> Noting the company's most significant overseas investment along the Belt and Road Initiative, GCL's president Zhu Gongshan said the lack of power supply in the regions has provided Chinese energy companies with an important opportunity to export technology, employment, and better energy mix abroad.[73]

Chinese state media appears to have ramped up other socio-economic significance of the deal by pointing out that "[a]nalysts said China and Indonesia are highly complementary and with many Chinese descendants in Indonesia, it is relatively easy for Chinese enterprises to invest in the

[70] Xinhua, "China-invested power plant starts construction in Indonesia to resolve electricity shortage" dated 25 Oct 2016 in the Xinhuanet website [downloaded on 25 Oct 2016], available at http://www.xinhuanet.com/english/2016-10/25/c_135780336.htm

[71] Xinhua, "China-invested power plant starts construction in Indonesia to resolve electricity shortage" dated 25 Oct 2016 in the Xinhuanet website [downloaded on 25 Oct 2016], available at http://www.xinhuanet.com/english/2016-10/25/c_135780336.htm

[72] Jakarta Globe, "China Invests $17b in Indonesian Power Project" dated 29 May 2013 in the Jakarta Globe website [downloaded on 1 Jan 2018], available at http://jakartaglobe.id/business/china-invests-17b-in-power-project/

[73] Zheng, Xin and Zou Shuo, "Coal-fired power project kicks off" dated 10 August 2017 in China Daily [downloaded on 10 August 2017], available at http://www.chinadaily.com.cn/business/2017-08/10/content_30402968.htm

country."[74] On the one hand, the presence of ethnic Indonesian Chinese whose communities hold tremendous economic power makes it easier for mainland Chinese investments to find partners in Indonesia. It is an advantageous relationship. On the other hand, such statements may be controversial in Indonesia that has a small overseas Chinese minority that historically bear the brunt of extremist nationalistic sentiments, sometimes viewed with suspicions and questioned about their loyalties to the Indonesian state. Media commentaries aside, the Chinese private sector is also keen to be a new entrant into West Kalimantan's power industry which is at present served by high-quality diesel generators from non-Chinese sources and higher-cost imported equipment from Malaysia, according to Chinese state media report and analysis.[75] Chinese sources quoting the Indonesian government figures dated December 2016 also argue that the archipelago is still lacking in power supply to its isolated regions with 2519 villages having no access to electricity.[76]

In November 2017, Indonesian Energy and Mineral Resources Minister Ignasius Jonan inked a deal with China's National Energy Administration Director Nur Bekri to enhance collaboration in the field of energy at the fifth Indonesia-China Energy Forum (ICEF V) in Jakarta and set up two working groups on oil, gas, and coal + renewable energy for electricity.[77] Chinese involvement in Indonesia is now comprehensive, from buying raw coal commodities to supplying power plants for their Indonesian clients. Chinese firms have started 35,000-MW electricity provision to Indonesia through two engineering, procurement, and construction (EPC) arrangements and an independent power producer (IPP) while Aluminum Corporation of China Ltd. (Chinalco) partnered with PT Aneka Tambang

[74] Jakarta Globe, "China Invests $17b in Indonesian Power Project" dated 29 May 2013 in the Jakarta Globe website [downloaded on 1 Jan 2018], available at http://jakartaglobe.id/business/china-invests-17b-in-power-project/

[75] Zheng, Xin and Zou Shuo, "Coal-fired power project kicks off" dated 10 August 2017 in China Daily [downloaded on 10 August 2017], available at http://www.chinadaily.com.cn/business/2017-08/10/content_30402968.htm

[76] Zheng, Xin and Zou Shuo, "Coal-fired power project kicks off" dated 10 August 2017 in China Daily [downloaded on 10 August 2017], available at http://www.chinadaily.com.cn/business/2017-08/10/content_30402968.htm

[77] The Jakarta Post, "Indonesia, China sign agreement on energy" dated 13 November 2017 in Jakarta Post [downloaded on 13 November 2017], available at http://www.thejakartapost.com/news/2017/11/13/indonesia-china-sign-agreement-on-energy.html

and PT Inalum to build a smelter in West Kalimantan in a deal that is worth at least US $1.5–$1.8 billion.[78]

If the Chinese investments in Indonesia work well, it could very well be touted as a new model of great power economic relations in the East Asian context, between the world's second largest economy and the world's largest Muslim country. Due to the scale and complexities of the deals, obstacles are unavoidable and they would be interesting case studies to see how both countries navigate through the issues. The Indonesian coal investments also lend credence to the argument that Chinese megaprojects in countries listed within its BRI have already started BRI-related projects even before BRI was up and running. In other words, China may derive valuable experiences in running these BRI-category public infrastructure projects and then use them as references and applications in other BRI projects. Thus, the Chinese-built and/or Chinese-invested coal plants in Indonesia are valuable as showpieces for the overall BRI.

The Case Studies of Other Northeast Asian States: Russia

Russia is a major source of coal energy commodities in Northeast Asia as the Putin administration stopped operations at non-productive coal mines to make energy supply more effective and to create industry models that can rival the most efficient coal industries globally.[79] Russian Siberian coal mines are close to Japan, especially found in West Siberia. According to statistics dated around 2014, Russia exported 110 million tons to Japan, the two fast-emerging economies of India and China as well as the European Union (EU) consumers, while in the same period of time, Japan took in 121 million tons of coal from Indonesia, Australia, Russia, the US and Canada.[80]

[78] The Jakarta Post, "Indonesia, China sign agreement on energy" dated 13 November 2017 in Jakarta Post [downloaded on 13 November 2017], available at http://www.thejakartapost.com/news/2017/11/13/indonesia-china-sign-agreement-on-energy.html

[79] Campi, Alicia, "The New Great Game in Northeast Asia: Potential Impact of Energy Mineral Development in Mongolia on China, Russia, Japan, and Korea" dated 2013 in the Asia-Pacific Policy Papers Series [downloaded on 1 Jan 2017], (Washington, DC: Edwin O. Reischauer Center for East Asian Studies), 2013, available at http://www.reischauercenter.org/en/wp-content/uploads/2013/06/RC-Monograph-2013-Campi_The-New-Great-Game-in-Northeast-Asia.pdf, p. 9.

[80] Enkhbold, Voroshilov, "Mongolia's Coking Coal Export Potentials in Northeast Asia" in The Association of Korean Economic Studies (AKES) website [downloaded on 1 Jan 2017], available at http://www.akes.or.kr/eng/papers(2014)/120.full.pdf, p. 8.

The Case Studies of Other Northeast Asian States: North Korea

Interestingly, North Korea has quite big reserves approximated at 12 GT and potentially undiscovered coal mines.[81] Its problems in coal extraction are manmade. Years of isolation, sanction, and self-reliance *juche* have not brought about rapid economic development or cutting-edge industrialization. Its low-cost labour potentially makes North Korean coal cheap but the payback for taking Pyongyang's resources is political repercussions. When the North Koreans tested a missile in early 2017, China turned back North Korean ships carrying coal for export. North Korea is under increasing international sanctions, even from its ally China, for its missile testing and nuclear programmes. Its recent test fire of Intercontinental Ballistic Missiles (ICBMs) and threats to attack Guam did not go down well with the international community. The US has promised tougher crackdown on Pyongyang's trade lifeline and financial activities while opening the door for talks. These sanctions and tougher stance against Pyongyang is likely to stifle North Korean industries (including coal mining) even further.

The Case Studies of Other Northeast Asian States: Mongolia

Mongolia is bordered by Russia and China, both of which exert tremendous economic influence on Mongolia as a supply source for coal and other minerals while China is fast emerging as a major consumer of coal in the region. As it is located in a land-locked region, consequently, the major Mongolian export items to Japan (a maritime nation) consist of mostly minerals and textiles.[82] Significantly, 43% of China's coal resources is now drawn from Mongolia (estimated total of 18,473.2 million tons of coal) and there are more than 5000 companies from China with a total investment of US $2.5 billion managing energy trade (in addition to PetroChina's US $1.4 billion).[83]

[81] Streets, David G., "Northeast Asia coal trading center set up in Dalian" dated 1 July 1991 in the GLOCOM website [downloaded on 1 Jan 2017], available at http://www.glocom.ac.jp/column/1991/07/energy_and_acid_rain_projectio.html, unpaginated.

[82] Enkhbold, Voroshilov, "Mongolia's Coking Coal Export Potentials in Northeast Asia" in The Association of Korean Economic Studies (AKES) website [downloaded on 1 Jan 2017], available at http://www.akes.or.kr/eng/papers(2014)/120.full.pdf, p. 3.

[83] Campi, Alicia, "The New Great Game in Northeast Asia: Potential Impact of Energy Mineral Development in Mongolia on China, Russia, Japan, and Korea" dated 2013 in the Asia-Pacific Policy Papers Series [downloaded on 1 Jan 2017], (Washington, DC: Edwin O. Reischauer

Looking Forward: Common Challenges Faced

The problems with tapping into Northeast Asian coal mines are typically political ones with geopolitical tensions in the region, including imposed economic sanctions, maritime disputes, and issues of historical memories. There are other common problems in using coal as well. For example, acid rains result from coal use in the region. Sulphur dioxide and nitrogen oxide are two by-products of coal use. They can cause respiratory problems and damage the ecosystem. Use of low-quality coal can worsen air pollution in the region. Densely populated cities found in the Northeast Asian region intensify the problem further as energy use is concentrated in those areas. Heavy industries in areas like Shenyang or other regions producing cement and steel also add on to pollution generated through coal use. In addition, domestic household consumption of energy including coal is likely to go up in expanding urban cities. Driving urbanization is the massive movements of people from the agricultural areas in the countryside to urban cities. This movement of people takes place on a massive scale in the context of China.

Trends in the Regional Coal Industry

Problems in coal use are well known. An example is acid rains from coal use in the region. Sulphur dioxide and nitrogen oxide are two by-products of coal use. They can cause respiratory problems and damage the ecosystem. Use of low-quality coal can worsen air pollution in the region. Densely populated cities found in the Northeast Asian region intensify the problem with the concentrated use of energy use. Heavy industries in areas like Shenyang or other regions producing cement and steel also add on to pollution generated through coal use. Domestic household consumption of energy including coal is likely to go up in expanding urban cities. Driving urbanization will drive massive people from the agricultural areas in the countryside to urban cities.

Like China, other Northeast Asian economies are also switching to cleaner fuels. Environmental pollution is also a major universal challenge that faces coal use in Northeast Asia and a common solution in the region

Center for East Asian Studies), 2013, available at http://www.reischauercenter.org/en/wp-content/uploads/2013/06/RC-Monograph-2013-Campi_The-New-Great-Game-in-Northeast-Asia.pdf, pp. 3 and 4.

has been to phase out smaller, less regulated, family owned (even illegal) coal plants that use obsolete technologies. The closure of smaller-scale coal-powered plants is a shift towards the utilization of larger plants. China is phasing out smaller-scale coal-fired plants in favour of large plants with capacities of 500,000 kW.[84] There are regional environmental cooperation initiatives amongst Northeast Asian states like China, Japan, Russia, Mongolia, and South Korea which have established formal diplomatic relations with each other after the Cold War. These countries realize the advantages of combining resources to tackle common transboundary and cross-border environmental problems and pollution (including those caused by coal use).

Future solutions to solve these challenges include the use of clean coal technology (CCT) or the use of scrubbers to remove the pollutants, but they add to the cost of using coal energy for power generation. The Great East Japan Earthquake 2011 has accelerated Japanese demand for coal since nuclear plants had to be shut down nationwide. Since 2011, only one nuclear power plant located in Kagoshima (in the southern Japanese island of Kyushu) has been turned back on. The current pro-nuclear Liberal Democratic Party (LDP) of Japan led by Prime Minister Shinzo Abe (a leader with staying power in Japan's political arena) and his cabinet/government at the point of this writing is keen to turn on other nuclear stations after the relevant safety inspection checks have been carried out.

Japanese CCT has been cited by the Turnbull government of Australia as a model for replacing their old coal power plants with cleaner versions.[85] The Isogo (in southern Tokyo) electricity generation plant is seen globally as the cleanest coal power plant using the criteria of emissions intensity.[86] Japan has 90 coal power plants in operation in 2017 and has intentions to add more plants, and this represents an export opportunity for Australia

[84] Fukushima, Atsushi, "Coal and Environmental Issues in Northeast Asia" dated March 2004 in The Institute of Energy Economics, Japan (IEEJ) website [downloaded on 1 Jan 2017], available at http://eneken.ieej.or.jp/en/data/pdf/242.pdf, p. 27.

[85] Curtis, Katina, "Look at Japan for clean coal power" dated 2 Feb 2017 in News Limited [downloaded on 2 Feb 2017], available at http://www.news.com.au/national/breaking-news/look-at-japan-for-clean-coal-power/news-story/784d9cb50af857bfa8cbc9f5bc90acc8

[86] Mealey, Rachel, "Japan spruiks 'highly efficient' coal-fired power plants as stop-gap measure to Australia's energy crisis" dated 1 June 2017 [downloaded on 1 June 2017], available at http://www.abc.net.au/news/2017-06-01/japan-spruiks-coal-fired-power-plants-to-australia/8577464

since the Japanese already purchase more than 40% of Australia thermal coal exports.[87] It is poised to purchase more with the expansion of Japanese coal plants. Coal itself can also be cleaned and the process lowers the sulphur and sodium contents. Like its carnation in seawater, sodium causes corrosion while sulphur contributes to acid rain formation.

Another solution to combat pollution issues like these is to conserve energy or use the energy efficiently to reduce wastage. These measures will lead to lower consumption of energy and reduce pollution as well. Substitution is another measure. Coal can be substituted by the much cleaner fossil fuel of natural gas and/or renewable energy resources like hydropower, wind, solar, and geothermal. According to a media report by a writer with the Australian Associated Press, renewable energy makes up 25% of Japanese power generation.[88] Japan has also contributed to a cleaner regional environment by providing training on environmental monitoring to shore up more reliable data, technological sharing, and promoting transboundary cooperation. Networks of experts have also been formed amongst Northeast Asian states to look into this issue. The major sources of coal energy for Japan remain the coal-rich regions of Canada, the US, and Australia, although both Japan and China are making more coal investments in Southeast Asia.

One of the primary targets for such investments in Southeast Asia is Indonesia. Chinese state-owned enterprises (SOEs) and government loans in particular team up strong to invest in Southeast Asian economies. While China is keen to shore up its credentials as the world's environmental power after the US backed off the Paris climate agreement and is actively adopting cleaner energy sources like renewables to lessen dependence on coal, its coal investments overseas have increased. At the point of this writing, Chinese financial institutions and enterprises are involved in a minimum of 79 coal electricity generation facility/projects adding up to 52 GW or more in electricity volume.[89] Such investment plans fit into

[87] Curtis, Katina, "Look at Japan for clean coal power" dated 2 Feb 2017 in News Limited [downloaded on 2 Feb 2017], available at http://www.news.com.au/national/breaking-news/look-at-japan-for-clean-coal-power/news-story/784d9cb50af857bfa8cbc9f5bc90acc8

[88] Curtis, Katina, "Look at Japan for clean coal power" dated 2 Feb 2017 in News Limited [downloaded on 2 Feb 2017], available at http://www.news.com.au/national/breaking-news/look-at-japan-for-clean-coal-power/news-story/784d9cb50af857bfa8cbc9f5bc90acc8

[89] Walker, Beth, "Chinese Investment Stokes Global Coal Growth" dated 24 September 2016 in The Diplomat [downloaded on 1 Jan 2017], available at http://thediplomat.com/2016/09/chinese-investment-stokes-global-coal-growth/

Southeast Asian energy priorities because electricity needs are poised to go up by 83% from 2011 to 2035 and a large number of Southeast Asian countries are still using coal-powered electricity generation facilities.[90]

The view that demand for coal and other fossil fuels will still increase in Southeast Asia is supported by International Energy Agency (IEA). Its study indicated that, although there are greater interest and use of renewable energy, the percentage of fossil fuels in Southeast Asia's energy mix will go up to approximately 80% by 2040.[91] A competing narrative, however, argues that Southeast Asia is following China's and, to a lesser extent, India's move away from coal energy in favour of cleaner energy resources like solar, nuclear, and natural gas. This school of thought argues that the main reasons that coal is still around are due to state policies that still favour coal in the region and abundant funding from Japan[92] (and China) but acknowledges that coal use (and its accompanying industry) is on the cusp of decline in the long run.

Concluding Remarks for the Chapter: Whither Northeast Asian Cooperation?

In the early days of Northeast Asian cooperation, China had already expressed through a Senior Official's Meeting (SOM) in 1993 that Beijing is interested in interacting/a dialogue with Japan and South Korea on topics of technology transfer and financial assistance mechanism building. China had proposed that the three countries embrace the "spirit of Agenda 21" to embark on a joint research project on clean coal combustion technology (CCCT).[93] The Chinese were less keen on joint

[90] Myllyvirta, Lauri, "Southeast Asia is planning 400 new coal power plants—what does that mean?" dated 13 January 2017 in the Energydesk Greenpeace website [downloaded on 13 Jan 2017], available at http://energydesk.greenpeace.org/2017/01/13/southeast-asia-coal-plans-health-japan-indonesia/

[91] International Energy Agency (IEA), "Emergence of Southeast Asia as energy giant carries risks, opportunities" dated 8 October 2015 in the IEA website [downloaded on 1 Jan 2017], available at https://www.iea.org/newsroom/news/2015/october/emergence-of-southeast-asia-as-energy-giant-carries-risks-opportunities.html

[92] Coca, Nithin, "Asia and the Fall of Coal" dated 22 June 2017 in The Diplomat website [downloaded on 22 June 2017], available at http://thediplomat.com/2017/06/asia-and-the-fall-of-coal/

[93] Taek-Whan Han, "Northeast Asia Environmental Cooperation: Progress and Prospects" dated 23 September 1994 in the Nautilus Institute website [downloaded on 1 January 2017], available at http://nautilus.org/trade-and-environment/northeast-asia-environmental-cooperation-progress-and-prospects-3/

research and monitoring systems for transboundary air pollution which are of greater interest to Korea and Japan.[94] However, cooperative initiatives are often affected by geopolitical rivalries, historical memories, and territorial disputes.

Working on common environmental problems including those caused by the use of coal fuel is useful for strengthening interdependence amongst Northeast Asian states since most of them are either highly industrialized (like Japan and South Korea) or are large emerging economies with maturing growth (e.g. China). Cooperation in such areas was however mired in political differences. Rapid industrial growth (and the accompanying urbanization) after China's economic reforms took off from the 1980s accentuated the use of coal for industrialization, heating, and other purposes, contributing to regional pollution. This was the same case for Japan during its rapid industrial growth in the 1960s (the "income-doubling" period).

Having discussed the historical evolution of the Japanese coal industry, its peak, and then closure to the gentrification of former domestic coal mines, and finally the Northeast Asian context, the next chapter will wrap up the volume with the intangible aspects of the industry, studying the gender perspectives, literary (poems, folk songs, films, and women's narratives) subtexts and artistic expressions of coal mining stakeholders/workers/family members throughout its industrial history. It adds the soft cultural perspective to the political economic, sociological, and historical perspectives in the previous chapters and highlights the human story in an otherwise quantitative interpretation of industrial development. The last chapter before the conclusion wraps up the story of the success, trials, and tribulations of the industry's history by analysing coal mine cultures in an intimate peek into the lifestyle choices, sense of aesthetics, leisurely entertainment of the coal mining actors, beyond just imagining them as agents of productivity and faceless individuals in a modern industrial apparatus.

[94] Taek-Whan Han, "Northeast Asia Environmental Cooperation: Progress and Prospects" dated 23 September 1994 in the Nautilus Institute website [downloaded on 1 January 2017], available at http://nautilus.org/trade-and-environment/northeast-asia-environmental-cooperation-progress-and-prospects-3/

Bibliography

Associated Press, "Chinese Coal Official Who Hid $30M Admits to Corruption" dated 29 December 2015 in the VOA website [downloaded on 1 January 2017], available at https://www.voanews.com/a/china-corruption/3122856.html

Bradshernov, Keith, "Despite Climate Change Vow, China Pushes to Dig More Coal" dated 29 November 2016 in the New York Times [downloaded on 1 January 2017], available at https://www.nytimes.com/2016/11/29/business/energy-environment/china-coal-climate-change.html?mcubz=1

Campi, Alicia, "The New Great Game in Northeast Asia: Potential Impact of Energy Mineral Development in Mongolia on China, Russia, Japan, and Korea" dated 2013 in the Asia-Pacific Policy Papers Series [downloaded on 1 Jan 2017], (Washington, DC: Edwin O. Reischauer Center for East Asian Studies), 2013, available at http://www.reischauercenter.org/en/wp-content/uploads/2013/06/RC-Monograph-2013-Campi_The-New-Great-Game-in-Northeast-Asia.pdf

Coca, Nithin, "Asia and the Fall of Coal" dated 22 June 2017 in The Diplomat website [downloaded on 22 June 2017], available at http://thediplomat.com/2017/06/asia-and-the-fall-of-coal/

Curtis, Katina, "Look at Japan for clean coal power" dated 2 Feb 2017 in News Limited [downloaded on 2 Feb 2017], available at http://www.news.com.au/national/breaking-news/look-at-japan-for-clean-coal-power/news-story/784d9cb50af857bfa8cbc9f5bc90acc8

Enkhbold, Voroshilov, "Mongolia's Coking Coal Export Potentials in Northeast Asia" in The Association of Korean Economic Studies (AKES) website [downloaded on 1 Jan 2017], available at http://www.akes.or.kr/eng/papers(2014)/120.full.pdf

Feng, Shirley, Yang Liu and Luna Lin, "China bans North Korean iron, lead, coal imports as part of U.N. sanctions" dated 14 August 2017 in The Washington Post [downloaded from https://www.washingtonpost.com/world/china-bans-north-korea-iron-lead-coal-imports-as-part-of-un-sanctions/2017/08/14/a0ce4cb0-80ca-11e7-82a4-920da1aeb507_story.html?utm_term=.2045009a70ea]

Fukushima, Atsushi, "Coal and Environmental Issues in Northeast Asia" dated March 2004 in The Institute of Energy Economics, Japan (IEEJ) website [downloaded on 1 January 2017], available at http://eneken.ieej.or.jp/en/data/pdf/242.pdf

Global Times, "Annual coal trade fair in Qinhuangdao" dated 27 November 2016 in the Global Times website [downloaded on 1 January 2017], available at http://www.globaltimes.cn/content/1020536.shtmlQinhuangdao

Hewett, Jennifer, "Japan has good reason to burn coal for electricity" dated 3 Feb 2017 in The Australian Financial Review website [downloaded on 3 Feb 2017], available at http://www.afr.com/business/mining/coal/japan-has-good-reason-to-burn-coal-for-electricity-20170202-gu4csz

International Energy Agency (IEA), "Emergence of Southeast Asia as energy giant carries risks, opportunities" dated 8 October 2015 in the IEA website [downloaded on 1 Jan 2017], available at https://www.iea.org/newsroom/news/2015/october/emergence-of-southeast-asia-as-energy-giant-carries-risks-opportunities.html

International Energy Agency (IEA), Key Coal Trends Excerpt from: Coal Information 2016 (Paris: IEA), 2016.

Jakarta Globe, "China Invests $17b in Indonesian Power Project" dated 29 May 2013 in the Jakarta Globe website [downloaded on 1 Jan 2018], available at http://jakartaglobe.id/business/china-invests-17b-in-power-project/

Lim, Tai Wei, EAI Background Brief No. 1329 – COAL AND ITS IMPORTANCE IN CHINA'S ENERGY MIX (Singapore: National University of Singapore EAI), 2018 (28 Feb 2018).

Lim, Tai Wei, "The State of Coal Industry in Northeast Asia" dated 10 August 2017 in IPP Review website [downloaded on 10 August 2017], available at http://ippreview.com/index.php/Blog/single/id/527.html

Lin, Alvin, "Understanding China's New Mandatory 58% Coal Cap Target" dated 17 March 2017 in Natural Resources Defence Council (NRDC) website [downloaded on 17 March 2017], available at https://www.nrdc.org/experts/alvin-lin/understanding-chinas-new-mandatory-58-coal-cap-target

McHugh, Babs, "Japanese government planning to build 45 new coal fired power stations to diversify supply" dated 1 February 2017 in the ABC website [downloaded on 1 Feb 2017], available at http://www.abc.net.au/news/rural/2017-01-31/japan-coal-power-plants/8224302

Mealey, Rachel, "Japan spruiks 'highly efficient' coal-fired power plants as stop-gap measure to Australia's energy crisis" dated 1 June 2017 [downloaded on 1 June 2017], available at http://www.abc.net.au/news/2017-06-01/japan-spruiks-coal-fired-power-plants-to-australia/8577464

Meidan, Michael, "Political Transition Will Override China's Policy Targets" dated 17 March 2017 in the Chatham House website [downloaded on 17 March 2017], available at https://www.chathamhouse.org/expert/comment/political-transition-will-override-china-s-policy-targets

Myllyvirta, Lauri, "Southeast Asia is planning 400 new coal power plants—what does that mean?" dated 13 January 2017 in the Energydesk Greenpeace website [downloaded on 13 Jan 2017], available at http://energydesk.greenpeace.org/2017/01/13/southeast-asia-coal-plans-health-japan-indonesia/

Reuters staff, "China says enforcing North Korea coal ban seriously, no violation" dated 21 April 2017 in Reuters [downloaded on 21 April 2017], available at http://www.reuters.com/article/us-storm-harvey/trump-to-visit-victims-of-unprecedented-floods-in-texas-and-louisiana-idUSKCN1BD074

Shepard, Wade, "If China Is So Committed To Renewable Energy, Why Are So Many New Coal Plants Being Built?" dated 8 July 2016 in Forbes.com [downloaded on 8 July 2016], available at https://www.forbes.com/sites/

wadeshepard/2016/07/08/if-china-is-so-committed-to-renewable-energy-why-are-so-many-new-coal-plants-being-built/#36dbe1235918

Streets, David G., "Energy and Acid Rain Projections for Northeast Asia" dated 1 July 1991 in the International University of Japan GLOCOM website [downloaded on 1 Jan 2017], available at http://www.glocom.ac.jp/column/1991/07/energy_and_acid_rain_projectio.html

Tabeta Shunsuke, "China moving to renew coal mining curbs" dated 23 February 2017 in Nikkei Asia [downloaded on 23 February 2017], available at https://asia.nikkei.com/Politics-Economy/Economy/China-moving-to-renew-coal-mining-curbs

Taek-Whan Han, "Northeast Asia Environmental Cooperation: Progress and Prospects" dated 23 September 1994 in the Nautilus Institute website [downloaded on 1 Jan 2017], available at http://nautilus.org/trade-and-environment/northeast-asia-environmental-cooperation-progress-and-prospects-3/

Tan, Huileng, "China's wild futures trading is opening up a major opportunity for exchanges abroad" dated 21 May 2017 in the CNBC.com website [downloaded on 21 May 2017], available at https://www.cnbc.com/2017/05/21/heres-how-commodity-exchanges-are-eyeing-chinas-wild-futures-trade.htm

The Editors of Encyclopædia Britannica, "Qinhuangdao (Alternative Title: Ch'in-huang-tao)" in Encyclopaedia Britannica [downloaded on 1 August 2017], available at https://www.britannica.com/place/Qinhuangdao China's coal imports up 33% to nearly 90 million tonnes in Q1

The Jakarta Post, "Indonesia, China sign agreement on energy" dated 13 November 2017 in Jakarta Post [downloaded on 13 November 2017], available at http://www.thejakartapost.com/news/2017/11/13/indonesia-china-sign-agreement-on-energy.html

US Energy Information Administration, "Chapter 4. Coal" dated 11 May 2016 in International Energy Outlook 2016 [downloaded on 1 Jan 2017], available at https://www.eia.gov/outlooks/ieo/coal.php

VOA, "Despite Renewable Energy Push, China Still Runs on Coal" dated 21 November 2011 in the VOA website [downloaded on 1 August 2017], available at https://www.voanews.com/a/despite-renewable-energy-push-china-still-runs-on-coal-134331173/168226.html

Walker, Beth, "Chinese Investment Stokes Global Coal Growth" dated 24 September 2016 in The Diplomat [downloaded on 1 Jan 2017], available at http://thediplomat.com/2016/09/chinese-investment-stokes-global-coal-growth/

Walsh, Bryan, "King Coal's Comeback" dated 11 June 2012 in Time (NY: Time Asia), 2012.

Xinhua, "China-invested power plant starts construction in Indonesia to resolve electricity shortage" dated 25 Oct 2016 in the Xinhuanet website [downloaded on 25 Oct 2016], available at http://www.xinhuanet.com/english/2016-10/25/c_135780336.htm

Zeng, Xiaolin, "Coal mining in China" dated 30 May 2017 in the IHS Fairplay: Maritime Shipping News website [downloaded on 30 May 2017], available at https://fairplay.ihs.com/commerce/article/4287086/china-s-coal-imports-up-33-to-nearly-90-million-tonnes-in-q1

Zhao, Qian and Roger Bradshaw (editors), "Dalian plays host to Northeast Asia Coal Trade Fair" dated 6 August 2013 in China Daily [downloaded on 1 January 2017], available at http://www.chinadaily.com.cn/m/dalian/2013-08/06/content_16874898.htmEXCHANGES

Zheng, Xin and Zou Shuo, "Coal-fired power project kicks off" dated 10 August 2017 in China Daily [downloaded on 10 August 2017], available at http://www.chinadaily.com.cn/business/2017-08/10/content_30402968.htm

Zhou, Xin, "Decline and fall: the broken dreams of a Chinese coal-mining city struggling to address industrial overcapacity" dated 30 May 2017 in the South China Morning Post [downloaded on 30 May 2017], available at http://www.scmp.com/news/china/economy/article/1935326/decline-and-fall-broken-dreams-chinese-coal-mining-city

CHAPTER 6

Coal Mining Subcultures: A Brief Cultural History of Hokkaido and Comparative Perspectives with Other Regions

In terms of tone and texture, this chapter stands out from the rest. It looks at coal miners not as faceless individuals or agents of productivity but as humans who have individual preferences in terms of lifestyle choices, ideas about aesthetics, sense of belonging/identity, and choice of leisurely entertainment. The setting for this chapter is based on Hokkaido but its culture is comparatively examined alongside other coal mining regions. The coal mines also produced some of the most prolific non-tangible cultures like folk songs where miners sang in solidarity and in times of hardship. Harsh as well as unstable conditions motivated the emergence of self-help communitarian cultures and encouraged the coal miners and their families to play hard in group-oriented entertainment cultures. These conditions may be the same case in other coal mines nationwide. The chapter ends off with how culture was curated, conserved, and commodified in present-day Hokkaido. With the demise of the coal mining industry, Hokkaido's former coal mining towns are transitioning to service sectors that offer economic activities like retailing, tourism, hospitality, and cultural experiences.

Genesis

The image of Ezo (the old name of Hokkaido) and Hokkaido (after settlement by main island Honshu Yamato clan *Wa* Japanese) is that of a pioneering wild north. The lifestyle was significantly different from the rest of

T. W. Lim et al., *Coal Mining Communities and Gentrification in Japan*, https://doi.org/10.1007/978-981-13-7220-9_6

Japan, with its wide, open spaces, romantic images of Mother Nature waiting to be tamed by humans and the hardships that come with frontier spaces. Cultures were borne out of frontier settlers instead of urbanites or townsfolk. The frontier spaces developed their own cultures forged by the baptism of fire of humans conquering Nature. Hokkaido was constantly in the process of being settled by organized groups of humans. In terms of material culture, obsidian was amongst the first materials adapted for civilization as pressure forged shiny volcanic stones that were picked up by humans who discovered that they can be sharpened to create pointed edges to kill prehistoric animals for food. This was the first glint of Hokkaido civilization as these obsidian stones can be used as both weapons and jewellery. Thus, the same stone that was weaponized was also used as Hokkaido's first material culture. Shaping jewellery became one of the earliest expressions of aesthetics by Hokkaido's earliest frontier civilizations. Out of the hunter-gatherers of the earliest civilizations in Hokkaido, the indigenous Ainu culture was born with colourful fabrics, consumption of bear meat and herrings, carved decorative items, and so on.

Arrival of the "Wajins" and the Emergence of Coal Mining Subcultures

In the 1600s, Hokkaido culture and civilization changed again when the indigenous Ainu warriors went to war with encroaching *Wajins* (Japanese Honshu main islanders, literally translated as the *Wa* people). The *Shamo* guns of the Honshu army overcame the poison arrows of the Ainu fighters and Ezo was subjugated by the Honshu main islanders. The territory was renamed "Hokkaido" and Ainu culture was gradually mixed with incoming main island Honshu-based *Wajin* Japanese culture. By the time of Taisho democracy (1912–1926) when Japan became a fully industrialized modern liberal democracy, Honshu-based *Wajin* culture was entrenched and pluralistically present in Hokkaido. Some examples include the Minenobu east umbrella dance from Tottori first performed in the Minenobu wilderness (modern-day Minenobuchohigashi). The dance, which won the "Hokkaido cultural properties protection service prize," involves swinging swords and coarse oiled paper (a form of lacquered technique) umbrella and, since 2000, Bibai City has labelled this dance as an "immaterial folk cultural asset."[1]

[1] Japan Travel and Tourism Association, "Minenobu east umbrella dance" dated 2014 in the Japan Travel and Tourism Association (Nihon Kankou) website [downloaded on 1 Jan 2018], available at http://www.nihon-kankou.or.jp.e.wp.hp.transer.com/hokkaido/012157/detail/01215be2220093652

Perhaps the most important example of Honshu migrant culture permeating Hokkaido is the lion dances. As Hokkaido developed, its natural resources including coal were exploited for Japan's fledgling industrialization process in the nineteenth century. Migration boomed between the 1890s and 1920s due to economic opportunities for Honshu main island Japanese migrants while the Hokkaido Development Commission's foreign expert American engineer Benjamin Lyman completed a survey to map out Hokkaido coal resources.[2] According to the British consular reports dated 1912, 1869 was the official year in which the Japanese government proactively encouraged main island Japanese migration to Hokkaido.[3] The flood of migrants pouring into Hokkaido's nascent natural resource industries like coal brought along important cultural assets that included highly entertaining lion dances, of which two genres stood out.[4]

The first example was the Tadoshi lion dance introduced by the Akaiwa family who resided in the contemporary Tadoshi area of Fukagawa and choreographed a lion dance for the Tadoshi Shrine (it was a transplanted version of the dance from the Akaiwa clan's hometown of Inotani, Kamitaira Village, Higashitonami-gun, Toyama).[5] The dance was conserved/curated by Fukagawa City as an "intangible cultural asset," alongside two other dance traditions: the *Shojo Jishi Godan Kuzushimai* and *Osamunai-cho Shojo Jishimai* dances. The second example is the Ashibetsu folk lion dance first performed by main island settlers who arrived in the present-day Honcho area of Atsubetsu from the Tonami area of Toyama in 1895, featuring a team of ten individuals navigating the lion in a large, dynamic, and artistic dance.[6] The dance was awarded the "intangible

[2] Akarenga, "Hokkaido's History, Culture and Nature" in Akarenga portal site of Hokkaido's History and Culture [downloaded on 1 Jan 2018], available at https://www.akarenga-h.jp/en/hokkaido/kaitaku/k-04/

[3] Foreign Office and the Board of Trade, "Coal Mining Industry in the Hokkaido" dated June 1912 in the Diplomatic and Consular Reports Miscellaneous Series No. 682 (UK: His Majesty's Stationery Office), 1912, available at https://digital.soas.ac.uk/content/LO/AC/00/00/66/00168/PDF.pdf, p. 3.

[4] Akarenga, "Hokkaido's History, Culture and Nature" in Akarenga portal site of Hokkaido's History and Culture [downloaded on 1 Jan 2018], available at https://www.akarenga-h.jp/en/hokkaido/kaitaku/k-04/

[5] Akarenga, "Hokkaido's History, Culture and Nature" in Akarenga portal site of Hokkaido's History and Culture [downloaded on 1 Jan 2018], available at https://www.akarenga-h.jp/en/hokkaido/kaitaku/k-04/

[6] Akarenga, "Hokkaido's History, Culture and Nature" in Akarenga portal site of Hokkaido's History and Culture [downloaded on 1 Jan 2018], available at https://www.akarenga-h.jp/en/hokkaido/kaitaku/k-04/

cultural asset" by Ashibetsu City, with its curation and conservation spearheaded by the Ashibetsu Lion Dance Preservation Society.

While main island culture permeated into Hokkaido indigenous culture, new indigenous subcultures emerged in the coal mines of Hokkaido. One example is Bibai's special brand of taiko drums known as "the drum which does not fly." This was a form of folk art circulated within the coal mines of Bibai and passed down from generations before they were inherited by today's youths interested to carry on such cultural traditions, and their powerful beats are often featured and used to stoke up the atmosphere in *matsuri* traditional festivals.[7] Another example is the annual Akabira summer fire festival found in Akabira City's Festival Square. The festival features the Himojiyaki on Saturday night where participants in red loincloths ascend the largest spoil tip in Japan, start a fire arranged in the Kanji character for "fire" while the Civic Fireworks Display is organized on the consecutive Sunday where 5000 fireworks are used.[8] These dance and *matsuri* traditions had a common feature. All of them were communitarian in nature and encouraged the mobilization of the local community for a specific purpose, something useful for massed agricultural, industrial, and large-scale resource extraction activities.

Evolving Culinary Culture: From No-Frills Grilled Herring to Modern Food Items

Out of its colourful civilizational changes and transformation, Hokkaido culture was born, a mix of indigenous native elements and Honshu Japanese culture that was itself hybridized with Western modernity. Existing literatures also include Korean migrant labourers and culture into the mix (especially the coal mines of Hokkaido and Chikuhō, Kyushu).[9] In the late nineteenth century, a new subculture was born when coal was discovered in Yubari in the age of industrialization and modernization.

[7] Japan Travel and Tourism Association, "Drum which does not fly in Bibai-shi" dated 2014 in the Japan Travel and Tourism Association (Nihon Kankou) website [downloaded on 1 Jan 2018], available at http://www.nihon-kankou.or.jp.e.wp.hp.transer.com/hokkaido/012157/detail/01215be2220093653

[8] Japan Travel and Tourism Association, "The 47th Akabira fire Festival" in the Japan Travel and Tourism Association Nihon Kankou website [downloaded on 1 Jan 2018], available at http://www.nihon-kankou.or.jp.e.wp.hp.transer.com/detail/01218ba2210113521

[9] Knotter, Ad, "Migration and ethnicity in coal field history (worldwide)" dated January 2013 in the Hypotheses.org website (Maastricht University), 2013, p. 1.

While coal mining was brutish and nasty, it stimulated colourful coal mining-related culture amongst the miners as they made merry in their lull and rest time to distract themselves from the harsh work environment. Perhaps one of the most important cultures in this aspect that appeal to the senses is culinary culture. When weather permitted, the Hokkaido coal mining families, especially the women, took out herring fishes and grilled them on outdoor portable stoves.[10]

The grilled herring-fish meal culture in the coal mines is now accompanied by rich Hokkaido culinary dishes in the contemporary period. In fact, Hokkaido culinary culture has become a major tourist attraction. For example, the Japan Travel and Tourism Association highlighted two specialty dishes for Bibai City. The first culinary item in this genre is the *torimeshi* of Nakamura that originated from the modern Meiji Period. A whole chicken is smashed apart and then marinated with strong soy sauce and their flavours then seep into sugared rice enveloped in flavouring oil.[11] The Association also recommended the Bibai Yakitori which is another Meiji-era food item. The Bibai Yakitori is touted as a "soul food" with chicken organs, meat, skin skewers that are so popular that it has spawned a Bibai Yakitori Association in Bibai in 2006 that participates in the National Yakitori Communication Meeting.[12] This indicates their high popularity and fame nationwide amongst consumers. They were iconic of the types of food brought over by main island Honshu migrants keen to participate in the economic development and natural extraction opportunities in Hokkaido.

When it comes to fresh foods, Yubari farmers grow some of the best musk melons in Japan. Not native to the region, these melons are carefully cultivated from seeds bought over from the Europeans. They are carefully cultivated in nets with the farmers' tender loving care. The melons are well

[10] Hoffman Michael, "How Hokkaido, born to be wild, was tamed" dated 2 March 2003 in Japan Times [downloaded on 1 Jan 2018], available at https://www.japantimes.co.jp/community/2003/03/02/general/the-great-north/#.W_9Np9szaUk

[11] Japan Travel and Tourism Association, "Torimeshi of Nakamura" dated 2014 in the Japan Travel and Tourism Association (Nihon Kankou) website [downloaded on 1 Jan 2018], available at http://www.nihon-kankou.or.jp.e.wp.hp.transer.com/hokkaido/012157/detail/01215fa2260096383

[12] Japan Travel and Tourism Association, "Bibai Yakitori" dated 2014 in the Japan Travel and Tourism Association (Nihon Kankou) website [downloaded on 1 Jan 2018], available at http://www.nihon-kankou.or.jp.e.wp.hp.transer.com/hokkaido/012157/detail/01215fa2260097461

known for their pristine condition with vein lines on the melons meandering perfectly on the skin. There are different grades to the melon and many are differentiated individually with the cultivator's photograph as a sense of pride for their top melons. Melon flavouring is mixed into ice cream, with the melon taste typically integrated into Hokkaido's famous dairy products. Besides ice cream, the essence of the melons is also mixed into melon soda, melon Pocky (famous Japanese brand for flavoured breadsticks), melon cakes, chocolates, candies, pastries, and so on. There was even a Yubari melon-themed play space for adult consumers with kids. In 1985, the Yubari melon castle became an event space for children's activities. Melon essence is also mixed into alcohol, liqueur, wines, and the like. In fact, Yubari melon products have become somewhat of an iconic status in Japan and they can also be found sold in premium snack outlets in Hong Kong and Singapore. In fact, from the author's observation study, a box of Yubari melon flavour Pocky costs US $30 per box in some specialty shops in Singapore. With greater popularity, it may be possible for prices to go down with time.

In 2014, the earliest ripened Yubari cantaloupe melon reached £15,000 at auction, becoming one of the most costly fruits ever sold in the open market while the cheapest ones cost £11, making their fast-ageing cultivators wealthy.[13] The melons have become the subject of popular culture with their likeness included in the images of the official Melon Bear mascot with its mean expression and athletic build.[14] Interestingly, the greatest threat to the Yubari melon and its associated products came from a newly built expressway that enables visitors to bypass Yubari, a potential negative development for the city and its industries. Musk melons are not just the preserve of Yubari alone in Hokkaido, as Mikasa City also has its own special melons, watermelon, and wine products that are equally reputable due to the meticulous efforts of the farmers and the naturally fertile soil of Mikasa City.[15]

[13] Hendy, Richard, "Yubari, Japan: a city learns how to die" dated 15 August 2014 in The Guardian [downloaded on 1 Jan 2018], available at https://www.theguardian.com/cities/2014/aug/15/yubari-japan-city-learns-die-lost-population-detroit

[14] Hendy, Richard, "Yubari, Japan: a city learns how to die" dated 15 August 2014 in The Guardian [downloaded on 1 Jan 2018], available at https://www.theguardian.com/cities/2014/aug/15/yubari-japan-city-learns-die-lost-population-detroit

[15] Mikasa City, "The City of Mikasa: a great place where you can enjoy!!" in Mikasa City website [downloaded on 1 Jan 2018], available at http://www.city.mikasa.hokkaido.jp/hotnews/detail/00001591.html

Non-tangible Cultures from the Coal Mines

Cuisines are a highly visible material culture in Bibai and other parts of Hokkaido. But the coal mines also produced some of the most prolific non-tangible cultures like folk songs where miners sang in solidarity and in times of hardship. Ryutaro Hattori is a leading critic of Japanese folk songs in the early 1960s, compiling albums of such songs for American audiences and contextualizing them using English textual explanations. In the albums, Hattori explained how his interest and project came about through interacting with Americans who were exposed to Japanese folk songs, including those that originated from the coal mines. In the early 1960s, he was approached by a large-scale American publishing firm to publish Japanese songs from a book that Hattori edited in a collection of Japanese folk songs and one of them included the "Coal Miner's Songs of Joban,"[16] a coal mining region in Northeast Japan on the main Honshu island closest to Hokkaido.

Hattori had to deal with foreigners who were exposed to Japanese folk songs for the first time. One of the common complaints he received from his American acquaintances was that the songs had homogenous melancholic sad tunes and, thus, they were not able to tell them apart, but Hattori also noted that when foreigners develop a keener sense of listening to such songs, they soon discover the elusive differences.[17] This inspired Hattori to curate some of the most common songs for introduction to American audiences. Hattori's study of Japanese folk songs, including those sung in coal mines, is important as they reflect the reception end of cultural products, especially foreign audiences. For cultural dissemination to be successful, both the creative producers as well as the reactions of audiences are crucial. While miners' folk songs resonate with local Japanese due to language and cultural affinity, a real test would be marketing these songs to foreign audiences who do not speak Japanese and share little resonance with the contents of the songs.

The common characteristics that coal mine folk songs share with other folk songs are that they are all arranged in double or quadruple time, on a

[16] Hattori, Ryutaro, "Traditional Folk Songs of Japan Selected and arranged by Ryutaro Hattori from his own collection Part Two: West Japan" dated 1961 (NYC: Folkways Records), 1961, unpaginated [also available online at https://folkways-media.si.edu/liner_notes/folkways/FW04534.pdf]

[17] Hattori, Ryutaro, "Traditional Folk Songs of Japan Selected and arranged by Ryutaro Hattori from his own collection Part Two: West Japan" dated 1961 (NYC: Folkways Records), 1961, unpaginated [also available online at https://folkways-media.si.edu/liner_notes/folkways/FW04534.pdf]

pentatonic scale, and passed from generation to generation and tended to be sung either without musical instruments or with simple instruments like samisen, *koto* (Japanese zither), or *shakuhachi* (Japanese flute).[18] The Joban coal miner songs' origins coincided with the onset of the modern Meiji era and specifically written by the Joban coal miners from Ibaragi Prefecture. The contents of the songs talked about the brutish and nasty life in the mines, including waking up early, descending deep into the pits, and the need to handle explosives and falling rocks. The Joban song also included moralistic themes like filial piety (working for parents' sake) and self-sacrifice (Japanese women becoming widows when their husbands are killed by falling rocks), but also rejoiced at the financial rewards when minerals are found.

Coal Miner's Songs of Joban
From six in the morn,
I work in the mine,
Carrying a lantern to and fro—
All for my parents' sake.

Three thousand feet down the mine
And you are in hell'
Ere long you'll lie among the dust
Of an abandoned mine.

You explode dynamite,
You'll find some fragments left over;
And the fragments will bring
Gold enough for you.

Listen, you young girls,
Wives of coalminers
Will become widows
The moment the rock fall.

(Source: Hattori, Ryutaro, "Traditional Folk Songs of Japan Selected and arranged by Ryutaro Hattori from his own collection Part Two: West Japan" dated 1961 (NYC: Folkways Records), 1961, unpaginated [also available online at https://folkways-media.si.edu/liner_notes/folkways/FW04534.pdf])

[18] Hattori, Ryutaro, "Traditional Folk Songs of Japan Selected and arranged by Ryutaro Hattori from his own collection Part Two: West Japan" dated 1961 (NYC: Folkways Records), 1961, unpaginated [also available online at https://folkways-media.si.edu/liner_notes/folkways/FW04534.pdf]

It may be possible to argue that the living conditions of the coal mines in the early Showa period may have inspired contemporary coal mining cultures. First, there was a state of lawlessness in some of these mines, which gives rise to a wild frontier culture, both psychologically and cognitively. In Chikuhō coal mines of Northern Kyushu for example, law enforcement did not reach into the mines; the state left the mines to the supervision of the mining firms and their security personnel.[19] Gender blind, these miners worked in teams of 12 or so workers that integrated many women into their numbers and they were highly mobile mine hopping in the 1930s to 1945, following the availability of jobs and workers' quarters.[20] Such harsh as well as unstable conditions motivated the emergence of self-help communitarian cultures and encouraged the coal miners and their families to play hard in group-oriented entertainment cultures. These conditions may be the same case in other coal mines nationwide. Indeed, in terms of lyrical materials, whether in the form of folk songs in Kyushu, Joban, Northeast Japan, or rhymes in Yubari, Hokkaido, Japan, a commonality appears to be bleak themes of deaths in coal mines. For example, visitors to Yubari point out a traditional rhyme from that region also dealt with the theme of death and dangerous working conditions at the shafts (including the detonation of dynamite). In fact, the rhyme is somewhat self-deprecatory highlighting the woes and underdeveloped hilly terrains of Yubari.

> Yuubari, kuubari, saka bakari,
> Dokan to kureba, shinu bakari
> Yubari's a town of woe, there's nothing but hills,
> When the shafts blow, there's nothing but death
> An old Yubari rhyme

(Source: Spike Japan, "Yubari: From the culture of coal to the cult of caramel" in Spike Japan [downloaded on 1 Jan 2018], available at https://spikejapan.wordpress.com/spike-hokkaido-2/yubari-from-the-culture-of-coal-to-the-cult-of-caramel/)

[19] Allen, Matthew, "Undermining the occupation: women coalminers in 1940s Japan" dated 2010 in University of Wollongong Research Online Faculty of Arts Papers (Archive) Portal Journal of Multidisciplinary International Studies, 7 (2) [downloaded on 1 Jan 2018], available at https://ro.uow.edu.au/cgi/viewcontent.cgi?referer=https://www.google.com.hk/&httpsredir=1&article=2088&context=artspapers, p. 5.

[20] Allen, Matthew, "Undermining the occupation: women coalminers in 1940s Japan" dated 2010 in University of Wollongong Research Online Faculty of Arts Papers (Archive) Portal Journal of Multidisciplinary International Studies, 7 (2) [downloaded on 1 Jan 2018], available at https://ro.uow.edu.au/cgi/viewcontent.cgi?referer=https://www.google.com.hk/&httpsredir=1&article=2088&context=artspapers, p. 5.

By the 1950s, scholars were already researching on organized cultural activities managed by coal mines' residents, including literary circles and formation of choirs and theatre, alongside printed media written on various topics of interest related to the coal mines.[21] They sometimes took on a left-leaning theme, narrating difficult working conditions to their comrades overseas. The theme of oppression coincided with worldwide labour unrest and socialist movements during this era. University intellectuals like Mizutamari Mayumi are researching and curating stories from former Hokkaido coal miners in Yubari, Bibai, and Utashinai who are connected emotionally with their hometowns found in the periphery of Japan.[22]

Gender and Coal Mining Culture

Embedded into coal mining culture is a genderized theme. Women were equal contributors to the evolution of coal mine community cultures. It is difficult leaving women out of the coal mining cultural narratives due to their ubiquitous integration with the physical landscape of coal mines themselves. Out of the coal mining villages (*tanju*), many women exposed to alienation, physical beating, mental domination, bullying by both genders, and multiple identities as constituents of households, interest groups, and other organized units.[23] All these factors shaped cultural experiences in the female perspective. Written works and photographic albums like Tajima's *Tanko Bijin* written in the year 2000 is an affective biography of female miners based on oral interviews with noir photos of these women.

Kyushu female miners enjoyed the most egalitarian treatment alongside their male peers in the coal mining industry. Hokkaido women were apparently protected from coal mining work to the furthest extent amongst coal mining communities in Japan. Thanks to the School of Oriental and African Studies (SOAS), University of London (UOL) Digital Collections,

[21] Mizutamari, Mayumi, "Lab.letters" dated 2018 in the Graduate School of Letters/ Faculty of Letters, Hokkaido University [downloaded on 1 Dec 2018], available at https://www.let.hokudai.ac.jp/en/staff/3-6-06/

[22] Mizutamari, Mayumi, "Lab.letters" dated 2018 in the Graduate School of Letters/ Faculty of Letters, Hokkaido University [downloaded on 1 Dec 2018], available at https://www.let.hokudai.ac.jp/en/staff/3-6-06/

[23] Allen, Matthew, "Undermining the occupation: women coalminers in 1940s Japan" dated 2010 in University of Wollongong Research Online Faculty of Arts Papers (Archive) Portal Journal of Multidisciplinary International Studies, 7 (2) [downloaded on 1 Jan 2018], available at https://ro.uow.edu.au/cgi/viewcontent.cgi?referer=https://www.google.com.hk/&httpsredir=1&article=2088&context=artspapers, pp. 6–7.

the Diplomatic and Consular Reports on Japan's Hokkaido coal mining industry in 1912 are available for scholarly research. Based on observations made by British consular diplomats, Hokkaido women were employed for surface and underground work as labourers, carriers, and coal dressers at Poronai, Ikushunbetsu, and Yubari No. 1 from at least the 1910s, much lighter work compared to their Kyushu counterparts who are involved in actual coal extraction, supported by the views that results are optimal when both genders work together.[24] Regardless of region, eventually, all Japanese women were disallowed from working in the coal mines. After the Pacific War defeat of Japan in 1945, the Occupation authorities eventually banned women from the coal mines, classifying those jobs as dirty and dangerous. It may be useful to recall that one of the reasons for female participation in the heavy industrial and resource extraction workforce in the 1930s and 1940s was war mobilization. Ironically, in times of peace, protecting women from danger was a major cause for the termination of female physical culture and henceforth became a supporting role, one that was equally important.

Regardless of their roles and their multiple identities in coal mine communities, women naturally coalesce and bonded closely with other women. In fact, coal miners providing support for one another and providing self-help were highlighted by Naoko Shimazaki in her scholarly works. She makes the important argument that coal miners provide support for one another through personal networks. Women naturally played an important role in such personal networks, as wives, mothers, and fellow female miners' solidarity. In other words, women were also cultural glue that bonded family households and communities together. Women's natural tendency to coalesce around family units correspond to the generally group-ist, communitarian, and team-oriented culture of coal mining communities, whether in terms of *matsuri* festivals, personal networks, mutual support, or/and lifestyle choices (e.g. communal outdoor herring cooking in Hokkaido). The next chapter will be utilizing a gendered lens to understand the socio-cultural and historical contexts and working conditions under which Japanese women miners laboured, and, consequently, to analyse points of resistance, if any, that women displayed within this environment.

[24] Foreign Office and the Board of Trade, "Coal Mining Industry in the Hokkaido" dated June 1912 in the Diplomatic and Consular Reports Miscellaneous Series No. 682 (UK: His Majesty's Stationery Office), 1912, available at https://digital.soas.ac.uk/content/LO/AC/00/00/66/00168/PDF.pdf, p. 22.

Gentrification and Visual Culture

With the demise of the coal mining industry, Hokkaido's former coal mining towns are transitioning to service sectors that offer economic activities like retailing, tourism, hospitality, and cultural experiences. Consequently, as former coal mining areas gentrify, some become museums, others become open spaces with outdoor galleries, and some innovate new uses for the city spaces, for example, hosting an international film festival. Yubari in Hokkaido was also well known for its visual culture fest as it hosts one of the most important film festivals in the world. Visitors to Yubari can often spot nostalgic move posters peppered all over town depicting celebrities and stars (such as James Dean and Audrey Hepburn) from different eras of Hollywood films. Former coal mining towns are also reviving traditional festivals as a means to attract visitors. For example, Mikasa City has preserved two major *matsuri* festivals known as the Ume or Japanese plum blossom festival in May, featuring a competition for spitting pickled plum seed and the *Hokkai-bon-odori* traditional dance from 13 to 15 August 2018 that features a well-known traditional folk song *Hokkai-bon-uta*.[25]

Arents Tom Jos Hendrik in fact argues that culture and state structures are influential on how local communities manage coal mine closures through mutual support and personal networking though many in Yubari chose to leave the city along with their cultural heritage and "the soul and the identity" of the city.[26] To arrest the cultural decline of the city, the Yubari authorities constructed the Historical Coal Mining Museum and Investment Park in 1984 and the Yubari International Fantastic Adventure Film Festival in 1989 to reconstruct a tourism-oriented culture but unknowingly entered into competition with other gentrified coal mining towns and contributed to the Showa-era construction bubble eventually leading to bankruptcy.[27] The Yubari case study pointed out the limitations

[25] Mikasa City, "The City of Mikasa: a great place where you can enjoy!!" in Mikasa City website [downloaded on 1 Jan 2018], available at http://www.city.mikasa.hokkaido.jp/hotnews/detail/00001591.html

[26] Hendrik, Arents Tom Jos, "Regeneration in Former Coal-Mining Regions: Yubari and the Campine Compared" in Hosei University website [downloaded on 1 Jan 2018], available at https://www.hosei-web.jp/chiiki/taikai/170315/t_03.pdf, unpaginated.

[27] Hendrik, Arents Tom Jos, "Regeneration in Former Coal-Mining Regions: Yubari and the Campine Compared" in Hosei University website [downloaded on 1 Jan 2018], available at https://www.hosei-web.jp/chiiki/taikai/170315/t_03.pdf, unpaginated.

of state-led reconstruction of declining coal mines. Yubari also hinted that it might not be an optimal solution to revive coal mining cultures and identities by simply injecting funding. Market forces and demographic forces are equally critically influential on the process. In fact, sometimes Nature's reclamation of former human habitation and industrial structures can represent opportunities for green tourism and environmental well-being for the surrounding areas. Productivity and economic revenues may not be the topmost priority in such cases.

Hokkaido's Bibai appears to have greater success by heading towards the cultural route instead of commodified culture like Yubari. Bibai has gone the way of becoming a town known for its sculptures, featuring the major works of Kan Yasuda in the Kan Yasuda Sculpture Art Museum Arte Piazza Bibai founded in 1992. Bibai-born Kan converted the old closed-down Sakae Elementary School gymnasium into a studio, a development that mirrored the decline of the population in Bibai from 90,000 to 20,000, prompting Kan to "make a wide space to open the hearts of the children."[28] This was a reuse of abandoned buildings due to the ageing population and a way to revive visitors and local resident participation in Yubari. Besides the immovable sculptures, the "Heart Carving Class" workshop is organized at Arte Piazza Bibai on a monthly basis, working with participants to carve the imagined shape for their hearts using natural stone as raw material to work on. The session is carried at the café of the Arte Piazza Bibai with choices of Italian white marble or Hokkaido soft stone as the raw material accompanied by a brief lesson before participant embark on an artistic journey in a serene and green surrounding embedded with inspiring Kan works.[29] Creativity is encouraged in this journey of discovery by the individual participants. Even culinary culture is integrated into the process where participants lunched at 12 noon on dishes (including the famous *torimeshi* chicken rice, local sweet corn, etc.) made by local cultivators using Bibai food items while utensils like chopstick rests are made of white marble. Admission for

[28] Japan Hoppers Editorial Department, "'Heart Carving Class' in Bibai, a town influenced by coal mining and sculpture" dated 27 September 2017 in the Japan Hoppers website [downloaded on 1 Jan 2018], available at https://www.japanhoppers.com/en/features/leisure/418/

[29] Japan Hoppers Editorial Department, "'Heart Carving Class' in Bibai, a town influenced by coal mining and sculpture" dated 27 September 2017 in the Japan Hoppers website [downloaded on 1 Jan 2018], available at https://www.japanhoppers.com/en/features/leisure/418/

the workshop is free.[30] From ancient civilizations interacting with main islander migrant culture alongside the emergence of a coal mining subculture, the disappearance of coal mines has stimulated nostalgia for the past, resulting in the commodification of curated culture. Perhaps, this is an optimal sustainable way of conserving cultures using market forces.

Bibliography

Akarenga, "Hokkaido's History, Culture and Nature" in Akarenga portal site of Hokkaido's History and Culture [downloaded on 1 Jan 2018], available at https://www.akarenga-h.jp/en/hokkaido/kaitaku/k-04/

Allen, Matthew, "Undermining the occupation: women coalminers in 1940s Japan" dated 2010 in University of Wollongong Research Online Faculty of Arts Papers (Archive) Portal Journal of Multidisciplinary International Studies, 7 (2) [downloaded on 1 Jan 2018], available at https://ro.uow.edu.au/cgi/viewcontent.cgi?referer=https://www.google.com.hk/&httpsredir=1&article=2088&context=artspapers, pp. 1–14.

Foreign Office and the Board of Trade, "Coal Mining Industry in the Hokkaido" dated June 1912 in the Diplomatic and Consular Reports Miscellaneous Series No. 682 (UK: His Majesty's Stationery Office), 1912, available at https://digital.soas.ac.uk/content/LO/AC/00/00/66/00168/PDF.pdf

Hattori, Ryutaro, "Traditional Folk Songs of Japan Selected and arranged by Ryutaro Hattori from his own collection Part Two: West Japan" dated 1961 (NYC: Folkways Records), 1961, unpaginated [also available online at https://folkways-media.si.edu/liner_notes/folkways/FW04534.pdf]

Hendrik, Arents Tom Jos, "Regeneration in Former Coal-Mining Regions: Yubari and the Campine Compared" in Hosei University website [downloaded on 1 Jan 2018], available at https://www.hosei-web.jp/chiiki/taikai/170315/t_03.pdf, unpaginated

Hendy, Richard, "Yubari, Japan: a city learns how to die" dated 15 August 2014 in The Guardian [downloaded on 1 Jan 2018], available at https://www.theguardian.com/cities/2014/aug/15/yubari-japan-city-learns-die-lost-population-detroit

Hoffman Michael, "How Hokkaido, born to be wild, was tamed" dated 2 March 2003 in Japan Times [downloaded on 1 Jan 2018], available at https://www.japantimes.co.jp/community/2003/03/02/general/the-great-north/#.W_9Np9szaUk

[30] Japan Hoppers Editorial Department, "'Heart Carving Class' in Bibai, a town influenced by coal mining and sculpture" dated 27 September 2017 in the Japan Hoppers website [downloaded on 1 Jan 2018], available at https://www.japanhoppers.com/en/features/leisure/418/

Japan Hoppers Editorial Department, "'Heart Carving Class' in Bibai, a town influenced by coal mining and sculpture" dated 27 September 2017 in the Japan Hoppers website [downloaded on 1 Jan 2018], available at https://www.japanhoppers.com/en/features/leisure/418/

Japan Travel and Tourism Association, "The 47th Akabira fire Festival" in the Japan Travel and Tourism Association Nihon Kankou website [downloaded on 1 Jan 2018], available at http://www.nihon-kankou.or.jp.e.wp.hp.transer.com/detail/01218ba2210113521

Japan Travel and Tourism Association, "Bibai Yakitori" dated 2014 in the Japan Travel and Tourism Association (Nihon Kankou) website [downloaded on 1 Jan 2018], available at http://www.nihon-kankou.or.jp.e.wp.hp.transer.com/hokkaido/012157/detail/01215fa2260097461

Japan Travel and Tourism Association, "Drum which does not fly in Bibai-shi" dated 2014 in the Japan Travel and Tourism Association (Nihon Kankou) website [downloaded on 1 Jan 2018], available at http://www.nihon-kankou.or.jp.e.wp.hp.transer.com/hokkaido/012157/detail/01215be2220093653

Japan Travel and Tourism Association, "Minenobu east umbrella dance" dated 2014 in the Japan Travel and Tourism Association (Nihon Kankou) website [downloaded on 1 Jan 2018], available at http://www.nihon-kankou.or.jp.e.wp.hp.transer.com/hokkaido/012157/detail/01215be2220093652

Japan Travel and Tourism Association, "Torimeshi of Nakamura" dated 2014 in the Japan Travel and Tourism Association (Nihon Kankou) website [downloaded on 1 Jan 2018], available at http://www.nihon-kankou.or.jp.e.wp.hp.transer.com/hokkaido/012157/detail/01215fa2260096383

Knotter, Ad, "Migration and ethnicity in coal field history (worldwide)" dated January 2013 in the Hypotheses.org website (Maastricht University), 2013, pp. 1–9.

Mikasa City, "The City of Mikasa: a great place where you can enjoy!!" in Mikasa City website [downloaded on 1 Jan 2018], available at http://www.city.mikasa.hokkaido.jp/hotnews/detail/00001591.html

Mizutamari, Mayumi, "Lab.letters" dated 2018 in the Graduate School of Letters/Faculty of Letters, Hokkaido University [downloaded on 1 Dec 2018], available at https://www.let.hokudai.ac.jp/en/staff/3-6-06/

Spike Japan, "Yubari: From the culture of coal to the cult of caramel" in Spike Japan [downloaded on 1 Jan 2018], available at https://spikejapan.wordpress.com/spike-hokkaido-2/yubari-from-the-culture-of-coal-to-the-cult-of-caramel/

CHAPTER 7

The World of the Female Miner in Japan: Sites of Compliance and Resistance

Introduction

History is gendered and the history of coal mining is no different.[1] The importance of a gendered history in revising and adding to official accounts or master narratives cannot be understated.[2] These other narratives that stand outside the official sphere of commonly accepted knowledge undermine but also enhance our apprehension of not only the conditions of the industry but how these conditions impacted specific actors, enabling us to arrive at differentiated conclusions. As such, gendered analyses of the mining industry—and other analyses that question and critique existing modes of understanding, for example, according to class or race, or make use of alternative sources of information like oral histories and diaries—are gaining traction in this field of scholarship.[3] The creation of the Gender History Association (*Jendā Shi Gakkai*) in

[1] Barbara Molony and Kathleen Uno, introduction to *Gendering Modern Japanese History*, eds. Barbara Molony and Kathleen Uno (Cambridge, MA: Harvard University Asia Center, 2005), 2.

[2] Jean-Francois Lyotard's grand narratives or master narratives were introduced in his 1979 publication *The Postmodern Condition: A Report on Knowledge*. The concept critiques dominant ideologies and forms of knowledge in order to understand how they are legitimized and sustained.

[3] Some examples include publications by Hideko Idegawa, Mikiso Hane, Sachiko Sone, Tatsuichi Horikiri, Regine Mathias, Matthew Allen, W. Donald Smith, Laurier Mercier, and Angela V. John.

T. W. Lim et al., *Coal Mining Communities and Gentrification in Japan*, https://doi.org/10.1007/978-981-13-7220-9_7

Japan in 2004, for example, demonstrates the increasing acceptance of gender analyses in academic discourse.[4]

The histories and stories of coal mining women in Japan have, for the most part, been a neglected area of study. Previously, most historical surveys or sociological studies of Japan's economic miracle, technological progress, and industrial growth and development have looked at the coal mining industry as almost synonymous with the male miner, choosing to discuss the coal industry from such a perspective, presuming, inaccurately, that gendered experiences can be universalized, or leaving women out of the discussion altogether.[5] Such a view of the industry negates gender-specific experiences and does not look at differentiated effects and affects within the industry. Assuming this position essentially erases gender-specific circumstances, which are integral to a deeper understanding of the contributions and experiences of both women and men within the industry. An even more detrimental consequence is the effacement of the female miner from discussions of the industry, rendering her invisible within history.

The aim of this chapter, then, is to utilize a gendered lens to understand the socio-cultural and historical contexts and working conditions under which Japanese women miners laboured, and, consequently, to analyse points of resistance, if any, that women displayed within this environment. Instead of seeing "[h]istorical assumptions about the thoroughly 'masculine' nature of mining" as limiting, Laurier Mercier and Jaclyn J. Gier discern, within this definition of mining history, the opportunity to explore issues like the challenges women may have presented to "existing inequalities," and to ask questions that elucidate the relationships that exist between history, culture and gender identities, roles, and ideologies.[6] By concentrating on social history and community relations and dynamics, gender relations can be better understood within their relevant contexts. The gaps that are created by official local histories that claim to exclude bias by leaving out subjective accounts of events effectively nullify the efforts and contributions of actors within history.[7]

[4] Vera Mackie, "Gender and Modernity in Japan's 'Long Twentieth Century,'" *Journal of Women's History* 25, no. 3 (Fall 2013): 62.

[5] Kuntala Lahiri-Dutt, "The Megaproject of Mining: A Feminist Critique," in *Engineering Earth*, ed. Stanley D. Brunn (Dordrecht: Springer, 2011), 329.

[6] Laurier Mercier and Jaclyn J. Gier, introduction to *Mining Women: Gender in the Development of a Global Industry, 1670–2005*, eds. Laurier Mercier and Jaclyn J. Gier (New York: Palgrave Macmillan), 3.

[7] Matthew Allen, *Undermining the Japanese Miracle: Work and Conflict in a Coalmining Community* (Cambridge: Cambridge University Press, 2009), 33.

The increasing acceptance of gender analyses is part of a greater change in the presentation and understanding of history, which places emphasis on the advancement of counter-narratives that showcase aspects of history that depart from or add to existing official records maintained by the state or large corporations. In the case of coal mining, these counter-narratives represent history by focusing on a specifically working-class narrative and its "gender dimension," which is set up as a contrast to the post-war image of Japan as a nation of the middle class.[8] The role women play in this counter-narrative is significant considering their roles in the propagation and archiving of personal histories and their roles as activists, all of which challenge mainstream historical records known to the public, and are part of a broader culture that draws attention to the people's history and local cultures, traditions, and art.[9]

As such, the analysis in this chapter will be largely dependent on sources that are not typically included in official records. These include oral or written histories or interviews given by women miners, miners' songs, Sakubei Yamamoto's illustrations and accompanying texts, and the presentation of women miners in a documentary film by filmmaker Hiroko Kumagai, which focuses on the Miike Mine. It is hoped that these sources will not only reveal how much agency and voice women possessed, and/or felt they possessed, within the prescribed circumstances, but will be able to fill in the gaps left by official accounts that fail to incorporate and explain subjective experiences within the often anonymous and seemingly objective records of history.

History and Background

The most important and productive coal mines in Japan were in northern Kyushu and Hokkaido. Due to the smaller numbers of women working underground in Hokkaido, however, the analysis in this chapter will focus on the female coal miners in northern Kyushu, especially the Chikuhō region, who because of the prevalence of small and medium mining operations were part of the mining workforce for a much longer length of time, even when faced with government restrictions governing the participation of women

[8] Dhiraj Kumar Nite, "Archives, Being and Representation: Studies into the Formation and Usage of Grassroots Archives," *Waseda RILAS Journal*, no. 6 (October 2018): 576.

[9] Chelsea Szendi Schieder, "From Coal Miner's Wife to Historical Actor: The Personal Archive of Matsuo Keiko," *Waseda RILAS Journal*, no. 6 (October 2018): 546.

and children in the mines later in the twentieth century.[10] The mode around which mining was organized in Kyushu, particularly in Chikuhō, also made the use of female and child labour more suitable than in Hokkaido, where the labour force was mainly made up of single male migrant miners.[11]

Coal was discovered in Chikuhō during the sixteenth century and was, at the time, primarily used by the feudal lords for domestic purposes.[12] For the peasantry, mining for coal during this period was mainly carried out in family units and was undertaken not as formal employment but for supplementary income during the seasonal slack in agrarian communities.[13] Some changes to this earlier arrangement were experienced during the latter half of the nineteenth century:

> The number of coal-mines increased, and a growing number of them were managed by early 'entrepreneurs' who … pursued the extraction of coal in a more systematic way: For more and more people, many of them migrant workers, who toured the coal districts alone or with their families, mining became a full-time occupation. Despite this, on the whole coal-mining was still regarded as a seasonal occupation for the unskilled labour force of the surrounding villages and this image changed only very gradually later on, when industrialisation led to a great expansion of coalmining in Japan.[14]

One important point the aforementioned quote highlights is the migrant labour that would characterize many mining communities. This is an important observation especially if we consider how miners and their families were often isolated from other communities surrounding the mines and therefore also seen, frequently, as outsiders who did not belong.[15]

[10] It must be stressed that Hokkaido also had women who worked underground, primarily as *atoyamas* (hauliers who transported coal after it had been extracted from the coal face). In fact, these female miners in Hokkaido may be said to be even more marginalized in terms of the lack of discussion and study about them.

[11] Sachiko Sone, "Japanese Coal Mining: Women Discovered," in *Women Miners in Developing Countries: Pit Women and Others*, eds. Kuntala Lahiri-Dutt and Martha Macintyre (New York: Routledge, 2006), 57.

[12] Allen, *Undermining the Japanese Miracle*, 54.

[13] Yutaka Nishinarita, "The Coal-Mining Industry," in *Technology Change and Female Labour in Japan*, ed. Masanori Nakamura (Tokyo: United Nations University Press, 1994), 60.

[14] Regine Mathias, "Female Labour in the Japanese Coal-Mining Industry," in *Japanese Women Working*, ed. Janet Hunter (London: Routledge, 1993), 100.

[15] Kuntala Lahiri-Dutt and Martha Macintyre, introduction to *Women Miners in Developing Countries: Pit Women and Others*, eds. Kuntala Lahiri-Dutt and Martha Macintyre (New York: Routledge, 2006), 7.

The outsider status of miners was also exacerbated by the relatively large numbers of ***burakumin*** who were employed in the mines.[16] This group of people are considered outcasts of society and were frequently discriminated against throughout the country even though they were racially considered Japanese. These factors would contribute to both the inward-looking mentality of many miners and the formation of a unique and separate culture that existed on the frontiers of society. This culture would function to both repress and emancipate, in some measure, female coal miners from the usual gender roles and identities demanded of Japanese women from other industries or from the middle class as we will discuss later in the chapter.

Due to the prevalence of the family unit working in the mines of Chikuhō, women miners, and children, were active underground in this region even before the modern mines of the nineteenth century mentioned above were in operation. They primarily worked in small mines known as badger burrows (*tanuki bori*) and even though these families mined only to supplement their income from agriculture, it was natural that as mining operations started to formalize that both men and women being put to work underground became a common occurrence in this region with women becoming indispensable as the production of coal was stepped up but mechanization lagged behind.[17]

As the interest in Western technology and know-how picked up during the beginning of the Meiji-era (1868–1912), and industrial development was stepped up to usher Japan into the "modern" age, coal became the natural choice to fuel the new industrial-oriented economy.[18] However, the growth in the industry was uneven, especially between small and medium mining companies and the large mines run by the *Zaibatsu* (big family-run financial corporations), and the industry as a whole suffered during certain periods like the post-war recession in the early 1920s, which led to the push for rationalization although this would proceed slowly and would mainly affect the larger mines.[19]

[16] Mikiso Hane, *Peasants, Rebels, Women and Outcastes: The Underside of Modern Japan*, 2nd ed. (Lanham: Rowman & Littlefield, 2003), 154.

[17] W. Donald Smith, "Sorting Coal and Pickling Cabbage: Korean Women in the Japanese Mining Industry," in *Gendering Modern Japanese History*, eds. Barbara Molony and Kathleen Uno (Cambridge, MA: Harvard University Asia Center, 2005), 395–6.

[18] Allen, *Undermining the Japanese Miracle*, 55.

[19] Mathias, "Female Labour," 100.

Notwithstanding the periods where demand for domestic coal dropped and stockpiling was common, the industry output dramatically increased in the first half of the twentieth century. This increase was mainly achieved through the rapid hiring of labour and the number of miners increased exponentially from 23,000 to 225,000 miners during the period 1894–1919.[20] This number, though partially reduced during the 1920s, later climbed to support the war effort, swelling from 140,000 miners in 1932 to 440,000 by 1945.[21] The number of women miners making up the numbers cannot be overstated. Before the 1920s, women made up nearly one-third of the mining workforce in both the large and medium and small-scale mines. Moreover, two-thirds of the women miners were working underground.[22] In fact, comparing the growth rates of male (104%) to female (147%) colliers in the decade between 1909 and 1919 demonstrates that "women's contribution to this expansion of the labour force, and thus to the large increase in coal production, was much greater than that of men."[23]

Although the mining industry has been integral in the transformation and growth of industrialized Japan, the presence of women within this industry has often been overlooked and they have been left out of most history textbooks even though coal mining was the "second largest occupational category for women" at the beginning of the twentieth century.[24] This is partly due to the "traditional definition of history in a patriarchal, male-defined world," which did not prioritize the recording of women's lives, resulting in a lack of sources from which to reconstruct or understand these women and their way of life.[25] The traditional mining unit organized around the family, the *tsuchi-kumi*, may also only call on women to relieve absent family members and women were therefore not formally acknowledged as employed workers underground.[26] This was also the case in other industrialized nations like the UK where male family members

[20] Mathias, "Female Labour," 100.

[21] Allen, *Undermining the Japanese Miracle*, 57.

[22] Sachiko Sone, "Coal Mining Women Speak Out: Economic Change and Women Miners of Chikuhō, Japan," in *Mining Women: Gender in the Development of a Global Industry, 1670 to 2005*, eds. Laurier Mercier and Jaclyn J. Gier (New York: Palgrave Macmillan), 154.

[23] Mathias, "Female Labour," 104.

[24] Sone, "Japanese Coal Mining," 51.

[25] Sone, "Japanese Coal Mining," 51.

[26] Kayoko Yoshida and Reiko Miyauchi, "Invisible Labour: A comparative Oral History of Women in Coal Mining Communities of Hokkaidō, Japan, and Montana, USA, 1890–1940," in *Mining Women: Gender in the Development of a Global Industry, 1670 to 2005*, eds. Laurier Mercier and Jaclyn J. Gier (New York: Palgrave Macmillan), 138.

utilized female relatives in mining work underground, but these women were not recorded in colliery accounts nor paid by the mining company.[27]

Another reason for the lack of attention paid to female coal miners can be attributed to the geographically remote areas in which the mines were located and the often reclusive nature of the miners, which often made the collection of information hard.[28] Hideko Idegawa who has personally approached former female miners to record personal stories and interviews attests to the difficulty surrounding such an activity:

> There were a number of problems at first, the biggest being that people were reluctant to talk about the difficult conditions in which they had lived, and continue to live in. Only after a long time was I accepted as someone who could be trusted.
>
> In the late 1950s and 1960s, when the small mines began to shut down, there was a lot of hardship and suffering in Kurate. ... During this period of extreme poverty, I decided to join the *Kuroi Hane* movement. This was a women's movement dedicated to helping the poverty-stricken miners and their families. ... I guess that it was my involvement with the *Kuroi Hane* that allowed people to recognise that I was a serious person.[29]

Bound by a common fate, the miners were close-knit not necessarily by choice but by circumstance. Idegawa, the outsider, then needed to prove her trustworthiness and sincerity before being admitted into confidence, illustrating the obstacles that stood in the way of not only information collection but, more importantly, access to authentic evidence or sources that could effectively shed light on the real plight and lives of female miners both during the golden age and the decline of the mining industry in Japan.

As Matthew Allen observes,

> History has become a commodity in the hands of a cynical and conservative government. Because there are few sources that refute official versions of history, or even modify accounts, the duopoly of information control would seem to be complete, and it could be assumed that mass perceptions of history have been or will be standardised. Those who *could* criticise official versions of Chikuhō history are gradually dying out, and with them will go the last oral histories of the region, seen from the bottom up as it were.[30]

[27] Angela V. John, *By the Sweat of Their Brow: Women Workers at Victorian Coal Mines* (New York: Routledge, 1980), 20.

[28] Nishinarita, "The Coal-Mining Industry," 59.

[29] Hideko Idegawa, interview by Matthew Allen, 1987, quoted in *Undermining the Japanese Miracle*, 52.

[30] Allen, *Undermining the Japanese Miracle*, 31.

To understand the fraught and often complex history of coal mining in Japan would require more than just the frequently whitewashed and sterile history offered by official sources or institutions which depict human beings as mere figures and statistics to be studied and categorized in the service of research, or as human interest elements that glorify the aims of industrialization and economic growth.[31] This is the reason the turn to alternative sources of history is deemed necessary especially in the excavation of information regarding overlooked subjects of history like the coal women of this chapter.

Coal Mining Women

Most Japanese were not well acquainted with the contributions and way of life of miners until the 1960s when an interest in local cultural history and events like the Miike Strike catapulted the mining industry into the consciousness of the masses.[32] Even with this interest, however, women miners were frequently relegated to the shadows as many considered their labour to be part of the family unit, which did not merit independent consideration:

> coal-mining at that time was carried out as a kind of family work, and must therefore be regarded as the extension to a new kind of side-line of the usual kind of family work in rural communities. Over the course of time, for many men and women this side-line turned into their main occupation. Nevertheless, it is important to bear in mind the fact that female labour in mining evolved within the framework of family work, and was therefore quite different from, for example, women's work in the textile mills.[33]

Unlike women working in the textile mills, women's labour in the coal mines were not considered independent contributions separate from the family unit. This is in spite of the fact that more than 5000 and 3000 women in Chikuhō worked as *saitanfu* (miners) and *teko* (miner's helpers) respectively, which meant that 70% of women labourers in Chikuhō were engaged in actual mining at the beginning of the twentieth century.[34] This

[31] See, for example, Matthew Allen's discussion of the presentation of history in the Tagawa museum in *Undermining the Japanese Miracle*.

[32] Sone, "Coal-Mining Women Speak Out," 154.

[33] Mathias, "Female Labour," 103.

[34] Sone, "Japanese Coal Mining," 63.

in turn affected the perception the public and scholars held towards the value of their work. Extensive studies of women's labour in the textile mills have been carried out in part because of the more formalized system of apprenticeship in operation at many mills and also because these women's work were seen to be individual contributions. This is in contrast to female coal miners, whose contributions are often discussed and studied as part of a team of workers rather than as individual contributions, downplaying their sense of agency and autonomy in the process and justifying, up till recently, the marginalization of their contributions to industrialization.

A family unit of coal miners was typically made up of at least one pair of male and female miners, usually a married couple or relatives, or siblings, or parent-child team. The male *sakiyama* or hewer would extract the coal while the female *atoyama* or haulier would transport the extracted coal to loading points where she would then load the coal into wheeled barrels or containers and push them to the surface. This arrangement encouraged the hiring of miners as a family unit especially in the earlier days of mining or in the small and medium mines where rationalization, mechanization, and methods of mining were not as far advanced as in the larger mines.

The hiring of family units was considered ideal within the mining industry. Consider the recruitment guidelines issued by the Mitsui Tagawa mine in 1906, which stated that ideal hires would be families which comprised a couple with a teenager, and an elder. The couple and child could work while the elder could keep house, ensuring maximum productivity.[35] Hiring both women and children effectively doubled the workforce in a labour-intensive industry at lower costs for the mining company since they were paid much less than men. Moreover, employing families was also seen as a way to ensure a continuing source of workers as many children tended to follow in their parents' footsteps, which provided a steady labour supply to the industry. This steady supply of labour was aided in part by the stabilizing effect hiring families had on the workforce. It was thought that recruiting couples and families could more successfully settle or fix what was essentially a mobile and migrant labour force. This mode of thought was supported by the turnover rates of single workers, which were typically two to three times higher than married workers.[36]

[35] W. Donald Burton, *Coal-Mining Women in Japan: Heavy Burdens* (New York: Routledge, 2014), 17.

[36] Mathias, "Female Labour," 104.

Because of this dependence on the family unit, early mining operations used the *naya seido* (subcontracting system) to hire and maintain control over the workers after recruitment. This system was suitable due to the labour-intensive nature of mining and the low level of skills needed to extract coal at this time. The *naya seido* was enforced by a *naya* boss (gang boss) who would recruit miners and maintain control over them in the company housing and mines largely through violence and intimidation. The gang bosses were overseers and saw fit to mete out "justice" or punishment according to circumstances. Mining companies depended on them to drive production and maintain or increase output when needed.[37] The *naya* system was gradually replaced by central mine management in the twentieth century when rationalization and mechanization changed how miners worked in the mines, rendering both the family unit and the *naya* boss less relevant to mine operations except in the smaller, less regulated, and frequently more dangerous mines in the region.

As technology progressed, and rationalization and increasing mechanization peaked in the 1920s, male and female miners were affected in disproportionate ways. The transition from the stall-and-pillar system to the longwall system of mining made women's work underground largely redundant, signalling the beginning of the end of family work in most mines.[38] Another development that contributed to the decrease in Japanese female miners was the increase in Korean miners who were put to work in the mines. Due to the implementation of the longwall method, mining teams were now made up of more workers in comparison to the two or three miners working in isolated teams previously. Moreover, the longwall method also eliminated the need for haulers, which was traditionally a job taken on by female coal miners.[39]

Apart from technology, policy changes also directly dictated the scope and viability of women's work in the mines. The restrictions to women's work that were applied to the mining industry in 1928 demonstrated clearly the government's need to control definitions of gender in terms of the state, social, and labour order, and connected the state's interests with the "protection" of women.[40] Consequently, many women either had to

[37] Sone, "Japanese Coal Mining," 55.

[38] Mathias, "Female Labour," 105.

[39] Nishinarita, "The Coal-Mining Industry," 90.

[40] Andrea Germer, Vera Mackie and Ulrike Wöhr, introduction to *Gender, Nation and State in Modern Japan*, eds. Andrea Germer, Vera Mackie and Ulrike Wöhr (New York: Routledge, 2014), 8.

take on work at the surface, which paid a lot less than underground work, or were dismissed altogether. Although ostensibly enacted for the protection of women, what these protective labour laws essentially accomplished was to reduce the wages of women miners, making it more difficult for families to sustain themselves, and also to push women to work in smaller collieries in which coal was still mined in traditional ways, which translated into lower pay, more dangerous working conditions, and worse living conditions.[41] Because all these consequences and effects coincided with the Great Depression, and rationalization and mechanization, this led many women to feel that the ban on underground work was not so much for their protection but for the smooth implementation of rationalization measures that would secure better profits for the mining companies.[42]

This sentiment that the protective labour laws were enacted not so much for women and minors as they were for the state and corporations must have been supported by the recalling of women into the mines when male labour was in shortage especially during the World War II when men were drafted. Volte-faces such as the above emphasized that gender-related boundaries and rules laid out in social and labour policy were to serve the state's economic and political needs and not, as overtly stated, exhibitions of real concerns about the welfare of women and minors.[43]

Even with the restrictions on women working underground, female coal miners worked at the coal face, often at smaller mines that did not adhere strictly to government laws or were granted exceptions due to the size of seams that needed traditional methods in order to extract coal. This continued even when the ban on women miners underground once again took effect after World War II. There are no official records on women miners after 1947, but oral accounts tell of women miners who continued working in small mines in both Chikuhō and Joban until the 1960s when most mines ceased operations in response to the shift in national energy policy from coal to oil.[44]

[41] Mathias, "Female Labour," 105.

[42] Regine Mathias, "The Nation at Work: Gendered Working Patterns in the Tashō and Shōwa Periods," in *Gender, Nation and State in Modern Japan*, eds. Andrea Germer, Vera Mackie and Ulrike Wöhr (New York: Routledge, 2014), 146.

[43] Mathias, "The Nation at Work," 147.

[44] Sone, "Japanese Coal Mining," 57.

The World of the Female Miner

The background presented in the previous section is integral to an understanding of the socio-cultural and historical contexts in which female miners lived and worked. Unlike the women who worked in the textile factories, or middle-class women who worked in urban centres, mining women existed within a certain nexus of relationships and systems that must be taken into account when analysing not only their lives but their capacity for autonomy and agency.

The world of mining communities is both familiar and unique in the way hierarchy and relationships are structured. Paternalism and the patriarchy were inherent in the way mining companies managed their workers, the way *naya* bosses supervised the miners under them, and the way family members interacted with one another. This essentially created a microcosm of traditional Japanese society and values within the isolated mining communities, but with unique features and concessions. Since most miners were migrants who left their villages to work and live in mining communities, the structure of control exercised by the *naya* bosses and/or the mining companies replaced what may be considered the extended social network labourers would be a part of if they had stayed in their hometowns or communities of origin:

> Miners who were migrant workers established their own communities, and within a mere half-century had created their own subculture, including a set of rituals and beliefs which had no parallels with coexisting communities such as farmers in rural areas and factory workers in the newly-established industrial centres. Yet, very much like the latter two groups, the miners' world was still constrained by feudal paternalism. This paternalism permeated the social fabric of imperial Japan from the lowest to the highest level of the social system. It was the feudal character signified by the subcontracting system (*naya seido*) that exploited miners and their women. Women, however, by virtue of gender, were doubly exploited: first by the subcontracting system, and then by men. Women at the turn of the 20th century had to work 20 hours a day, including labouring underground for 12 to 16 hours, and aboveground in the domestic sphere.[45]

The distinctive culture in mining communities created a different cultural space in which female miners existed and enacted their desires and agency. Although still subject to discrimination, abuse, assault, alienation, and

[45] Sachiko Sone, "The Reversible World of Japanese Coalmining Women," *The Australian Journal of Anthropology*, 18, no. 2 (2007): 212.

domination, the tight-knit bonds within the community were forged out of mutual recognition and the need for survival, and miners' interactions with one another extended beyond the underground world in which they worked, impacting all areas of their lives in the isolated communities in which they subsisted.[46] Summarized under the expression *Ichizan Ikka*, which translates to "one mine, one family," this motto exemplified the paternalism entrenched in coal companies.[47] Due to the geographically remote areas in which mines were located, and because most workers both lived and worked within the company's confines, miners developed a sense of solidarity with one another, replacing genuine familial bonds with the quasi-family bonds borne out of proximity and circumstance. Because the family unit was the primary working unit in many mines especially before rationalization, women had to grapple with both paternalism and patriarchal systems and attitudes simultaneously. This placed women in subordinate roles and statuses both within the family and within the company, effectively limiting their choices and agency.

One way in which the demands of patriarchy exerted itself was in the form of the working family unit. Women whose husbands had absconded or who were widowed were usually evicted from company housing and would not be able to secure work. The necessity of a male working partner thus placed women at the mercy of the *naya* system which dictated as necessary the attachment to a man, and therefore to a gang, in order to secure employment. Women had to hastily remarry—often with a *naya* boss playing matchmaker—for the sake of economic survival, particularly if they had children.[48]

The structures of paternalism and patriarchal attitudes and systems described above, however, were not always incompatible with women and their desire for self-worth and agency. This constituted the uniqueness of the socio-cultural network within mining communities:

> In the more remote areas of Chikuhō, in Northern Kyūshū, lay smaller, rougher, and more notorious coalmines, which to all intents and purposes were beyond the discourse of the state—which includes physical, ethical, legal, psychological factors, and control over the narratives of belonging—that is, coalminers and their families lived in a culture that bore little resemblance to that of city dwellers at the time. Police had no authority in these

[46] Allen, "Undermining the Occupation."
[47] Yoshida and Miyauchi, "Invisible Labour," 137.
[48] Burton, *Coal-Mining Women in Japan*, 109.

> mines or the nearby communities, governments turned a blind eye to mining activities, and many small mining towns were dominated by the control the companies and their enforcers extended in to the homes of the workers.[49]

The isolation of mining communities contributed to the somewhat independent system and culture that arose from the coming together of labourers and their employers. Unable and unwilling to police these rough communities that employed their own *yakuza* enforcers, the government all but relinquished control in these mining communities. This was simultaneously a blessing and curse for both male and female miners as they were not bonded to the strictures and expectations of society in the same way the urban population might have been. This proved advantageous in certain situations where rigidity would have been expected. For example, the usual practice of parents arranging marriages for their children seems to have been less successful in mining communities where young people who sought partners through love-matches were not as concerned about superstitions associated with the practice and suffered only minor social sanctions. They were also not as concerned with status, assets, or reputation since these could be said to be minimal or non-existent in the mining community, whose members were frequently regarded as being on the lowest ranks of Japanese society.[50]

The "narratives of belonging" miners ascribed to, therefore, were very much a unique product of economic and social influences. This was especially so for women who perceived, through their work, an avenue for agency and personhood. The often schizophrenic environment within mining communities with regard to expectations of behaviour and gender roles was both disconcerting and pressuring, and liberating at the same time. Pulled in different directions by the expectations generated by the patriarchy as well as economic necessity that very often subverted or undermined the patriarchy itself put the women in positions in which they experienced and resented the existing system of discrimination and/or control but were also offered opportunities to assert their autonomy and agency within the family unit or mining community, or in their perceptions of selfhood.

[49] Matthew Allen, "Undermining the Occupation: Women Coalminers in 1940s Japan," *PORTAL Journal of Multidisciplinary International Studies* 7, no. 2 (July 2010): 4.

[50] Burton, *Coal-Mining Women in Japan*, 105–6.

The value of the liminal space women occupied within mining communities cannot be overstated. This threshold within which women could both manipulate and be manipulated by the system gave these female miners a certain amount of leeway in which to exert their selfhood. Consider the ideology of the "good wife, wise mother" (*ryōsai kenbo*) which was used to define women's position and relationship to the state in modern Japan. The ideology framed women's relationship to the state as one that included the following: (1) a reproductive role—women were expected to give birth to and raise productive citizens; (2) a cultural role—women were inheritors of national culture and cultural identity and were expected to transmit these to their children; and (3) a symbolic role—women were represented as the (fragile) core of the nation that men had to protect.[51] Mining women were not beholden to the expectations that the ideology would create in the urban areas and did not overtly perform these roles. This would turn out to be a form of resistance in itself as the inhabitation of the roles circumscribed by the state was construed as acceptance of a gender identity that was defined as second class with respect to men.[52]

The importance of work, or more accurately, the value of work, therefore, in contributing to a more equitable status between the genders within the peasantry cannot be overstated. In the feudal period, there was greater equality between couples of the peasantry as compared to the warrior class precisely because the wife worked as hard as her husband and her contributions to the family wage and subsistence were seen as crucial and valuable.[53] Although this did not mean that the wife had the same autonomy and authority when it came to decisions regarding the family, she was regarded as a central figure within the family. However, as the ideology of *ryōsai kenbo* shows, women, regardless of their economic contributions, were the primary caregivers of the family and children, a role that males were not expected to fulfil.

This situation, though recognizing the contributions of the female miner as necessary and valuable, also placed additional burdens on the woman as opposed to the man. Expectations about gender roles dictated that women both had to work hard to contribute to the family *and* had to

[51] Mae Michiko, "The Resistance of Kanno Sugako and Kaneko Fumiko," trans. Leonie Stickland, in *Gender, Nation and State in Modern Japan*, eds. Andrea Germer, Vera Mackie and Ulrike Wöhr (New York: Routledge, 2014), 74–5.

[52] Michiko, "Resistance," 75.

[53] Hane, *Peasants, Rebels, Women and Outcastes*, 79.

fulfil her domestic duties which revolved around the house and children, in effect doubling (or more) the work and obligations a woman had to complete.[54] This naturally created pressures and obligations that women had to endure and manage that the men did not. Many mining men also, apparently, indulged in past-times like gambling and drinking, often impairing their ability to contribute to the family income.[55]

The woman's ability to work and sustain the family was crucial to any mining family. This engendered pride in many of the women as they not only contributed a substantial proportion of the family income but were also able to ensure their husbands turned up for work instead of going drinking and gambling because they worked in teams with them. Moreover, many female miners who had worked underground expressed that they had worked as hard as, or even harder, than their male counterparts. In fact, the income generated by these women was not a supplement to the family income but constituted a substantial part of it.[56] Women's underground work paid about 20–30% less than male miners, placing their earning power about 70% higher than female factory workers and approximately double that of the female coal sorters.[57] However, the trade-offs remain: dangerous working conditions, a relatively higher standard of living in mining communities, and high rates of accidents and illnesses were continual threats to the health and livelihood of these miners.

The mining community offered, through its unique ecosystem of relationships and structure and work expectations, a more open division of labour than was commonly conceived in other industries. At least until World War I, gender-related boundaries in coal mining remained rather fluid and was reflected in the relatively small gap in wage differences between the genders. Although males earned more than females, this seemed to be an attribute of the nature of work carried out rather than premised on gender.[58]

Female miners within a mining community transgressed typical gender expectations that were more pronounced and more rigidly adhered to in the urban workplace partly because of the physical nature of their work.

[54] Sone, "Japanese Coal Mining," 64.
[55] Burton, *Japanese Coal-Mining Women*, 111.
[56] Mathias, "The Nation at Work," 144.
[57] Smith, "Sorting Coal and Pickling Cabbage," 403.
[58] Mathias, "The Nation at Work," 143.

Take for instance the discussion below that compares the pouring of tea in a corporate environment with the mention of physical labour peasant women like female miners would be familiar with:

> "About this thing, we always speak of it in terms of equality between men and women. But, you see, women pour tea. If we have workplace equality, then women would even have to chop the firewood. That is harsh, but tea-pouring is women's work." [The female speaker] held an urban-centric understanding of the gendered division of work: in rural Japan, chopping firewood was women's work.[59]

What is underscored in the aforementioned utterance is the female (urban) speaker's gendered perceptions of what is considered women's work. In fact, the assumptions behind what constitutes women's work are so entrenched that even when questioned directly, the speaker did not feel that external factors like management personnel or company culture had decided these limitations and definitions of gendered work for employees. The speaker maintained that the action of pouring and serving tea was a natural expression stemming from the female employees themselves. This is starkly contrasted with the speaker's mention of chopping firewood, which she does not consider as appropriate for females because of the association with physical exertion. However, as has been explained previously, female miners and farmers (many of them inhabiting both roles) accepted within their scope of duties the discharge of manual labour and would not have felt a dissonance here. The physical nature of their work underground segregated them from "typical" female occupations and gender expectations, which exposed them to discrimination, but also offered a different space within which they could negotiate their identities and relationships—through work and wages—within their communities that were, to a certain extent, more fluid than other contexts would have allowed.

What is revealed through this analysis, therefore, is the fine line female miners tread between being economically valuable and economically dependent. Although their salaries were a necessity to the family and were a source of pride and self-confidence, their ability to earn an income was also oftentimes dictated by a patriarchal system that classified them as subordinate to male miners. Consequently, the self-confidence expressed by female miners was frequently dependent on a male figure in the first place,

[59] Christopher Gerteis, *Gender Struggles: Wage-Earning Women and Male-Dominated Unions in Postwar Japan* (Cambridge, MA: Harvard University Asia Center, 2009), 126.

who had to sanction their work in the mines. This complicates the perceived agency expressed by these women, which is more complex than a straightforward assertion of resistance through the value of their work, or their ability to pull their weight alongside their male counterparts.

Regardless of these limitations, many women who worked in the mines were, of their own volition, professional and committed to the job, though the necessity of work was also encouraged by the need for survival and the need to feed their families. Many women also picked up the slack when their husbands were unwilling or unable to work or contribute as efficiently to the family income, with some becoming main breadwinners for their families.[60] This, again, subverted the usual expectations and ideal of the male breadwinner.

In many mining societies and communities throughout the world, the masculine nature of mining is well documented. This was largely due to the physical and dangerous nature of mining work. The representation of mining as masculine served to legitimize the interests of male miners within the ideologies presented and embodied by the state, companies, and the family structure: the ideology of *ryōsai kenbo*, the overt paternalism demonstrated within company policies, and the behaviour of management, and the *naya seido*, for example, validated these masculine ideals as the norm. These systems and structures exerted influence over women's and men's consciousness, affirming men's sense of masculinity and superiority, and women's marginality and fragility.[61] This mode of thinking persisted in spite of the many economic realities—some of which have been described earlier in this chapter—that contradicted this ideal. As rationalization and mechanization were carried out and women were progressively phased out from underground work, the industry of mining became even more exclusively associated with men, relegating female miners to the forgotten spaces of history.[62]

But the marginalization of female miners need not continue. As mentioned earlier in the chapter, the masculine ideal of mining can be a fruitful area of research into the challenges that women posed to prevailing ideologies. Although marginalized within official records, these women can be known through other sources of information beyond statistics and figures, or the narrow focus on economic realities and modernization. Rather than

[60] Allen, "Undermining the Occupation," 6.
[61] Mercier and Gier, introduction to *Mining Women*, 5.
[62] Mercier and Gier, introduction to *Mining Women*, 5.

fixating on "objective" and impersonal accounts of the economic miracle in Japan, scholars can turn to other modes of historical accounts in order to flesh out these women on whom the coal mining industry has depended.

Looking Beyond: Possible Sites of Resistance?

Turning to other sources of history to understand the lives of women miners and the challenges they come up against proves to be a fruitful endeavour given the rich personal histories contained in alternative sources. The sources referred to in this section include mining songs, Sakubei Yamamoto's drawings and accompanying texts, the stories behind clothes worn during the period, Hiroko Kumagai's documentary film *Echoes from the Miike Mine*, and personal stories shared by female miners or coal miners' wives.[63]

One of the distinctive characteristics of mining communities is the departure from certain social mores for women in terms of attire, the expected behaviours and roles of women within state and home, and the representation of women with regard to femininity. Due to the female miner's transgressions—intentional or not—of these categories, roles, or boundaries, she becomes a liminal subject that attempts to refuse definition by external sources, which may, at various times and occasions, label her derogatively as masculine or present her as a denatured object because of her perceived violations of commonly held assumptions and accepted limits of gender roles.[64]

The following account by Keiko Matsuo, a coal miner's wife whose husband had previously worked in the Miike Mines and whose family had been involved in mining for three generations, describes how she perceived female miners and how mining is necessarily a gendered experience:

> The women who worked in the mines were so bold! They'd come back from the bath in the summer not wearing underpants, just a loincloth wrapped around their waist, and a wet towel draped over their shoulders, just tossed on, their breasts swinging in the breeze. They would stride along, that's how it seemed to me. And nobody thought anything of it.
>
> In the mine, their breasts would swell with milk when they were breastfeeding. "It was painful," they said. "I had to express the milk and throw it away. It made me so sad." They'd tell me stories like that. And I understood how they felt.

[63] Not all personal stories shared by female miners featured in this chapter have been acquired from primary sources due to the unavailability of certain publications in English.

[64] Lahiri-Dutt and Macintyre, introduction to *Women Miners in Developing Countries*, 7.

> In each 5-unit row house, there was one sink where we all did our wash. At the sink, when I was still breastfeeding my child, the baby cries, your breasts swell. I'd say, "Ah, she's crying," and run home, and sure enough, the baby would be up and crying.
>
> So you know, the working day back then was 8 hours or 10 hours, but there are things about it that only women understood.[65]

Matsuo's observations are significant for the emphasis they give to the female experience within mining that is frequently left out of sociological, economic, or historical studies of mining communities. She does not focus on the economic output or value of the women per se but attributes value to particular traits that she identifies as part of the unique experience that women undergo or have to endure within mining communities and as part of their mining work. Describing the female miners she sees as "bold," her admiration at their attire (or lack of), and the use of "stride" to convey a certain nonchalance, confidence, and even challenge in the way they walked convey her positive perceptions of these women who work in the mines. As she relates her account, her desire to associate herself with them and claim membership within their ranks is also important as it demonstrates how she regards them and their behaviour, attitudes, or even values as worthy of emulation. What is most important in her account, however, is the recognition that "there are things about [the mining experience] that only women understood," which is a clear indication of the need for a gendered analysis of the mining industry that can differentiate and therefore provide more nuanced understandings of the relationships between different historical actors, and industry and history.

This is why "narratives of belonging" are important. The habitus that the female miners inhabit and live contain and exhibit certain characteristics that then manipulate, necessitate, or sanction certain types of behaviours that allow these women to both adhere to and rebel against strictures levelled against them. Both the environment and the actors within it act upon one another within a culture that becomes an illustration of these very relationships that are played out.[66]

Of particular relevance in this context is the frontier mentality and environment that distinguishes many mining communities. Borne out of a

[65] *Echoes from the Miike Mine*, directed by Hiroko Kumagai (2005; Omuta, Fukuoka: SIGLO, Ltd., 2008). DVD.

[66] Pierre Bourdieu, *In Other Words: Essays Towards a Reflexive Sociology* (Cambridge: Polity Press, 1990).

number of factors including the geographically isolated locations of mines and therefore mining communities, the migratory nature of many miners and mining families, and the hands-off approach the government adopted with regard to many small and medium mining enterprises, the culture exhibited within these communities eschewed tenets of mainstream society, frequently subordinating social mores and expectations to profit and capital concerns.[67] The institutionalized violence within Japanese mines also constituted an integral part of the culture within these communities, subjecting both men and women to abuse and punitive measures that attempted to curtail any real agency on the part of the miners.[68]

The frontier mentality within mining communities facilitated the undermining or circumvention of labour laws with the tacit consent of both the labourers and the small-medium mining companies running operations. Minors, females and males alike, who falsely declared their ages to begin working in the mines before they reached the age of 16 were already a common occurrence when the Mining Law of 1905 or the Factory Law of 1911 were passed.[69] Moreover, mining companies frequently ignored directives concerning safety procedures, labour laws, or production targets.[70] Hence when the labour laws prohibiting women and children underground were passed in 1928, women who, by necessity, needed to work at the coal face, and mining companies, who considered these women valuable human resources, deliberately deceived the authorities through false declarations on employment rosters so as to prevent a disruption in the labour supply.[71]

These false declarations by women which were sanctioned by the mining companies could be seen as a form of resistance to state ideologies and demands, but understanding these actions as independent oppositions, that is, opposition by the individual for her self-interest, is also made difficult by the culture of control inherent within the industry that narrowed the scope of action and hence independence of women. The act of resistance in this instance can simultaneously be comprehended as being enacted with the "approval" of the culture within which these women

[67] Allen, "Undermining the Occupation," 11.

[68] Gill Burke, "Women Miners: Here and There, Now and Then," in *Women Miners in Developing Countries: Pit Women and Others*, eds. Kuntala Lahiri-Dutt and Martha Macintyre (New York: Routledge, 2006), 46–7.

[69] Burton, *Coal-Mining Women in Japan*, 99.

[70] Allen, *Undermining the Japanese Miracle*.

[71] Allen, "Undermining the Occupation," 10.

lived and worked, rather than as acts of resistance not only with regard to the state but also in the face of restrictions within these female miners' immediate environment.

Whether one considers the above an unequivocal act of resistance, it cannot be denied that the mining culture did afford the women manoeuvring space, which enabled them to articulate some form of personal agency within the physical and psychological confines they were in. One form of agency would be female miners who chose not to register their relationships as marriages, opting instead to remain as common-law wives (*naien no tsuma*) in order to retain their freedom. Being able to support themselves and their children through their work and wages, these women did not feel the need to subject themselves to the institution of marriage.[72]

Regardless of whether couples registered their relationships, women who lived with their male partners had to contend with a variety of experiences that only affected women. One of these was the burden of domestic duties mentioned earlier in the chapter. The lyrics of a mining song recorded in one of Sakubei Yamamoto's illustrations show the unequal burdens borne between genders:

> *Ishi wa chonkan demo jikan sae tateba,*
> *agarya nigo han ga udemakuri.*
> *Gotton!*
>
> Even if I can mine no more than a tub of coal today,
> 450 millilitres of sake will be waiting for me
> when I return home after my working hours are done.
> Gotton (Clang)! (Interjected chant)[73]

The lyrics showcase a male miner's perspective, highlighting the lack of household responsibilities for a male miner, and are recorded within an illustration that shows a woman preparing dinner while the man drinks sake after a long day at work. These domestic expectations and gender roles were common and contributed to the long hours of waged and unwaged work women were expected to shoulder in combination within

[72] Burton, *Coal-Mining Women in Japan*, 109. Do note that not all women chose this route as very often; two incomes (husband and wife team) were needed to support a family.

[73] Sakubei Yamamoto, "*Mukashi no Yama no Onna #14 (Tanju no Seikatsu; Fufu no Shokuji)* [Women at Coal Pits (Yama) in the Old Days #17 (Life in Row Houses; Meals of a Husband and Wife)] 1958–1963," trans. Nathan Johndro, accessed October 24, 2018, http://www.y-sakubei.com/world_appl_e/gallery.html?mode=view&pnum=143&cat=0

the peasantry. So although waged work was a source of pride for women, unwaged work often exerted its toll on them as the demands of both domestic and "professional" duties competed for attention.[74]

This competition between waged and unwaged work, especially in the specific realm of childcare, was partially "resolved" by having the whole family go into the mines for work. It was not uncommon for women to be carrying their babies on their backs while working in the mines or for older children to work with their parents within the mines, frequently also with a younger sibling in tow.[75] Older siblings could also be called on to keep house and take care of their younger siblings above ground with the mother coming up at intervals to feed infants if needed, and if possible.[76] Consider the following lyrics:

> *Nanatsu yatsu kara kantera sagete konai sagaru mo oya no bachi. Gotton!*
> It's because of my parents' fault that I had to work underground with a lamp in hand when I was seven or eight. Gotton (Clang)! (Interjected chant).[77]
>
> *Okete (normally okite) meshi kue konkon soete,*
> *konai sagaru mo taberu tame (oya no bachi [in red ink]).*
> *Gotton!*
>
> Get up and have rice with pickled radish,
> because we go into the pit to earn our meals (for our parents' fault).
> Gotton (Clang)! (Interjected chant).[78]

One of the consequences of these practices above was the lack of education experienced by many children born into mining families. They either did not manage to attend school regularly and dropped out eventually or took longer than usual to complete elementary school.[79] Mining children's early exposure to mining work and lack of education, as well as the financial

[74] Yoshida and Miyauchi, "Invisible Labour," 142.

[75] Hane, *Peasants, Rebels, Women and Outcastes*, 233.

[76] Sone, "Coal Mining Women Speak Out," 160.

[77] Sakubei Yamamoto, "*Mukashi Yama no Onna #12 (Kantera o Sagete Nyuko Suru Boshi)* [Women at Coal Pits (Yama) in the Old Days #12 (Mother and Children Entering the Pit with Lamps in Hands)] 1958–1963," trans. Nathan Johndro, accessed October 24, 2018, http://www.y-sakubei.com/world_appl_e/gallery.html?mode=view&pnum=134&cat=0

[78] Sakubei Yamamoto, "*Meiji Chuki Asa no Kofu* [Miners Having Breakfast in the Mid-Meiji Era (1868–1912)] October 1967," trans. Nathan Johndro, accessed October 24, 2018, http://www.y-sakubei.com/world_appl_e/gallery.html?mode=view&pnum=1&cat=11

[79] Sone, "Coal Mining Women Speak Out," 159.

needs of their families who were frequently in debt to *naya* bosses, were reasons contributing to the continued supply of labour in the mines. Poor families could not afford the luxury of allowing their children to continue their education beyond elementary school—if they even managed to complete this stage of education—and it was also not uncommon for peasants to send their single daughters off to work in the mines in order to sustain the family and supplement the income derived from farming.[80]

The female miner, therefore, not only had a responsibility to contribute to the family income but was intimately involved and invested in the well-being of her family. She was, one could say, the pillar holding the family unit together. Whether men absconded, were drafted into the military, or owed gambling debts, women were usually the ones who bore the brunt of the responsibility of taking care of the family, whether financially or otherwise:

> *Tate ba shakusen suware ba yachin,*
> *ayumu sugata ga hichiya yuki.*
> *Gotton!*
>
> When she stands, she is up to her ears in debt.
> When she sits, she is worrying about how to pay the rent.
> Judging by the way she walks, she is going to the pawnshop.
> Gotton (Clang)! (Interjected chant)[81]

Or as this set of lyrics illustrates sarcastically:

> *Musume yorokobe kondo no muko wa*
> *shigoto de kirai de sake ga suki.*
> *Gotton!*
>
> You are so lucky, daughter!
> Your new husband doesn't like to work, but likes to drink.
> Gotton (Clang)! (Interjected chant)[82]

[80] Tatsuichi Horikiri, *The Stories Clothes Tell: Voices of Working-Class Japan*, ed. and trans. Rieko Wagoner (London: Rowman & Littlefield, 2016), 61.

[81] Sakubei Yamamoto, "*Itcho Kiriha* [Single Coalface] June 1966," trans. Nathan Johndro, accessed October 24, 2018, http://www.y-sakubei.com/world_appl_e/gallery.html?mode=view&pnum=13&cat=13

[82] Sakubei Yamamoto, "*Meiji, Taisho, Showa Ki-kozumi (Fufu de no Sagyo)* [Wooden Cribbing in the Meiji (1868–1912), Taisho (1912–1926), and Showa (1926–1989) Eras (Work by a Husband and Wife)] 1964–1967," trans. Nathan Johndro, accessed October 24, 2018, http://www.y-sakubei.com/world_appl_e/gallery.html?mode=view&pnum=11&cat=5

Taking examples such as the above into account, it is not surprising that women were a more stable supply of labour in the mines than men in general. They are depicted in these songs as shouldering burdens created by their circumstances. Characterized both as victims and as problem-solvers, these women straddle the fine line between dependence and agency in these examples. They may have had these burdens to bear and may have been more reactive than proactive in their circumstances, but their commitment to work in these situations not only contributed to the family income but also helped companies keep up coal production, which was a vital role especially during the years when coal production was needed for the war effort.

Beyond stabilizing the highly mobile workforce, however, women also sometimes exercised independence in situations where their husbands or partners proved to be weak or incapable of supporting their families or working hard. A female miner who ended up with a *sakiyama* who only cared to drink, like the lyrics describe above, may choose to leave the wastrel for another more skilful miner or may simply "take up" with another *sakiyama* underground in order to earn her keep.[83] These women were usually capable *atoyama* who were skilled at their jobs and displayed physical prowess that may prove to be daunting, even to the men:

Onna nagara mo Takiyasha-hime wa
Shichiju goriki gama no jutsu.
Gotton!

Though she is a woman,
Takiyasha-hime can do the work of 75 people,
with her toad magic.
Gotton (Clang)! (Interjected chant)[84]

Kunya kunya otoko wa hanetobasareru,
noru ni norarenu hi no kuruma.
Gotton!

[83] Yoshida and Miyauchi, "Invisible Labour," 140.

[84] Sakubei Yamamoto, "*Mukashi Yama no Onna #17: Makitate de Kujibiki (Hako-tori no Kujibiki; Onna no Kamigata)* [Women at Coal Pits in the Old Days #17: Lottery Draw at the Landing (Lottery Draw for Empty Mine Cars; Women's Hair Style)] 1958–1963," trans. Nathan Johndro, accessed October 24, 2018, http://www.y-sakubei.com/world_appl_e/gallery.html?mode=view&pnum=80&cat=0

Soft and weak men will be ignored by her,
because she has such a furious temper,
like a fiery vehicle that no man can ride.
Gotton (Clang)! (Interjected chant)[85]

As these songs show, a woman who possessed strength and was hard-working at the coal face was simultaneously venerated as a being with "supernatural" abilities and as an anomaly that did not conform to the usual expectations and standards associated with the female gender. Both sets of lyrics, although ostensibly showing the physical prowess and abilities of the female miner, also exclude the figure of the female from ordinary human relations: she is either associated with folklore and "toad magic" or shown unable to be subdued (with sexual connotations) by males because "no man can ride" her. The female figure in these songs stand in opposition to other female perspectives and voices featured in other songs like the "Setto Song," or other mining songs in which the female singer pines for her lover who works in the mines and takes on significantly more feminine attributes, placing her firmly within the more traditional characterizations of the female in keeping with the ideology of *ryōsai kenbo*.[86]

The analysis of the lyrics featured above demonstrates the denaturing of female miners mentioned earlier, which firmly place them outside the circle of ordinary discourse, designating them as "others." But being excluded, in this case, from traditional conceptions of the female gender is not a loss. Their status as "others" imbues these female figures with power precisely because they stand outside of ordinary discourse and therefore possess the power to ignore, subvert, resist, and potentially overturn these definitions imposed on them. Women who displayed expertise at mining were called "goddesses of the pit," and there were even women who worked as *sakiyama*, directly competing against men and providing a source of pride and a positive role model to other women in the community.[87] Their sense of equal worth in relation to male miners underground

[85] Sakubei Yamamoto, "*Mukashi no Yamabito (Yufu no Kiridashi)* [People at Coal Pits (Yama) in the Old Days (Brave Woman Miner Cutting Coal Alone)] 1964–1967," trans. Nathan Johndro, accessed October 24, 2018, http://www.y-sakubei.com/world_appl_e/gallery.html?mode=view&pnum=5&cat=13

[86] Refer *to Kawaradake kara Mioroseba – Tankou Bushi no Genryu* [Looking Down from Kawaradake Mountain – The Origin of the Coal-mine Songs] for examples of these mining songs.

[87] Yoshida and Miyauchi, "Invisible Labour," 141.

was validated by their comparable hard work and coal production.[88] By existing within what Sachiko Sone terms the "reversible world" of coal mining, these women were able to exercise volition and impose their will—to a certain extent—on their external surroundings:

> There can be little doubt that for most women life underground, far from being a liberating experience, was one of unrelieved daily toil. But the reversible world ... reveals the capacity of women to redefine their role in society when the opportunity arises and circumstances demand it.[89]

Conclusion

The contributions of female miners within the Japanese coal mining industry may have been largely left out of official historical records, but there have been concerted attempts to revive the discussion and to restore these figures to their rightful place in the socio-cultural and historical fabric of the nation. Scholars in the field not only have conscientiously paid attention to alternative and forgotten sources of information in this area of analysis but have also studied and encouraged grassroots movements in coal mining regions that reposition ordinary human beings as historical actors. These historical actors capitalize on their direct experiences of the coal mining industry to enlighten the public through personal archives, the Internet, exhibitions, talks, interviews, and activism so as to shed light on the dark and forgotten arenas of history. These alternative records are important for consideration because they do not always align themselves with official ideology and narratives, enabling a fuller and more complex picture of industrialization to emerge.[90]

In the enthusiasm surrounding the recovery and celebration of local histories, traditions, customs, and culture, the public should not just celebrate the outward and obvious demonstrations of culture like the Tagawa Coalmine Festival or the United Nations Educational, Scientific and Cultural Organization sites, but ought to fully understand the stories and implications behind these celebratory displays. The incorporation of

[88] Hane, *Peasants, Rebels, Women and Outcastes*, 235.

[89] Sone, "Reversible World," 219.

[90] Refer, for example, to discussions by scholars (e.g. Takashi Miyamoto, Dhiraj Kumar Nite, and Chelsea Szendi Schieder) regarding the public archive in Japan and how the Internet and technology have aided activists and historical actors to share their narratives and to raise awareness.

personal accounts of residents living on Ikeshima Island in a commemorative publication focusing on its mining history is one such example.[91] Like this publication, discussions surrounding these subjects must move beyond specific and specialized circles into the public arena, where these narratives rightfully belong.

Bibliography

Allen, Matthew. *Undermining the Japanese Miracle: Work and Conflict in a Coalmining Community*. Cambridge: Cambridge University Press, 2009.

———. "Undermining the Occupation: Women Coalminers in 1940s Japan." PORTAL Journal of Multidisciplinary International Studies 7, no. 2 (July 2010): 1–14.

Bayliss, Jeffrey Paul. *On the Margins of Empire: Buraku and Korean Identity in Prewar and Wartime Japan*. Cambridge, MA: Harvard University Asia Center, 2013.

Bourdieu, Pierre. *In Other Words: Essays Towards a Reflexive Sociology*. Cambridge: Polity Press, 1990.

Burke, Gill. "Women Miners: Here and There, Now and Then." In *Women Miners in Developing Countries: Pit Women and Others*, edited by Kuntala Lahiri-Dutt and Martha Macintyre, 25–50. New York: Routledge, 2006.

Burton, W. Donald. *Coal-Mining Women in Japan: Heavy Burdens*. New York: Routledge, 2014.

Germer, Andrea, Vera Mackie and Ulrike Wöhr. "Introduction: Gender, Nation and State in Modern Japan." In *Gender, Nation and State in Modern Japan*, edited by Andrea Germer, Vera Mackie and Ulrike Wöhr, 1–24. New York: Routledge, 2014.

Gerteis, Christopher. *Gender Struggles: Wage-Earning Women and Male-Dominated Unions in Postwar Japan*. Cambridge, MA: Harvard University Asia Center, 2009.

Gier, Jaclyn J. and Laurie Mercier. "Introduction." In *Mining Women: Gender in the Development of a Global Industry, 1670 to 2005*, edited by Jaclyn J. Gier and Laurie Mercier, 1–10. New York: Palgrave Macmillan, 2006.

Gluck, Carol. *Japan's Modern Myths: Ideology in the Late Meiji Period*. Princeton, NJ: Princeton University Press, 1985.

Hane, Mikiso. *Peasants, Rebels, Women and Outcastes: The Underside of Modern Japan* (2nd ed.). Lanham: Rowman & Littlefield, 1982.

[91] Account of personal life on Ikeshima Island by Rie Miyahara in Hisaki Kurosawa, *Ikeshima Zenkei Ritoh no <Ikuukan>* [A Complete View of Ikeshima: The "Strange" Atmosphere of the Remote Island], ed. Isamu Sekiguchi (Tokyo: Masataka Shiomi, Sansai Books, 2017), 140–143.

Horikiri, Tatsuichi. *The Stories Clothes Tell: Voices of Working-Class Japan*. Edited and Translated by Rieko Wagoner. London: Rowman & Littlefield, 2016.

Hughes, David W. *Traditional Folk Song in Modern Japan: Sources, Sentiment and Society*. Folkestone: Global Oriental, 2008.

Idegawa, Hideko. Interview by Matthew Allen, 1987. Quoted in Matthew Allen, *Undermining the Japanese Miracle: Work and Conflict in a Coalmining Community*. Cambridge: Cambridge University Press, 2009.

John, Angela V. *By the Sweat of Their Brow: Women Workers at Victorian Coal Mines*. New York: Routledge, 1980.

Kawaradake kara Mioroseba – Tankou Bushi no Genryu [Looking Down from Kawaradake Mountain – The Origin of the Coal-mine Songs]. Released 19 January 2002. Columbia Music Entertainment. Compact Disc.

Kumagai, Hiroko, dir. *Miike – Owaranai Yama no Monogatari* [Echoes from the Miike Mine]. 2005; Omuta, Fukuoka: SIGLO, Ltd., 2008. DVD.

Kurosawa, Hisaki. *Ikeshima Zenkei Ritoh no <Ikuukan>* [A Complete View of Ikeshima: The "Strange" Atmosphere of the Remote Island], edited by Isamu Sekiguchi. Tokyo: Masataka Shiomi, Sansai Books, 2017.

Lahiri-Dutt, Kuntala. "The Feminisation of Mining." *Geography Compass* 9, no. 9 (September 2015): 523–541. https://doi.org/10.1111/gec3.12229.

———. "The Megaproject of Mining: A Feminist Critique." In *Engineering Earth: The Impacts of Megaengineering Projects* (Vol. 1), edited by Stanley D. Brunn, 329–351. Dordrecht: Springer Science and Business Media B.V., 2011. https://doi.org/10.1007/978-90-481-9920-4_20.

Lahiri-Dutt, Kuntala and Gill Burke. "Gender Mainstreaming in Asian Mining: A Development Perspective." In *Gendering the Field: Towards Sustainable Livelihoods for Mining Communities*, edited by Kuntala Lahiri-Dutt, 213–230. Canberra: Australian National University E Press, 2011. https://doi.org/10.22459/GF.03.2011.

Lahiri-Dutt, Kuntala and Martha Macintyre, "Introduction: Where Life is in the Pits (and Elsewhere) and Gendered." In *Women Miners in Developing Countries: Pit Women and Others*, edited by Kuntala Lahiri-Dutt and Martha Macintyre, 1–22. New York: Routledge, 2006.

Mackie, Vera. "Gender and Modernity in Japan's 'Long Twentieth Century.'" *Journal of Women's History* 25, no. 3 (Fall 2013): 62–91. Project MUSE.

Mathias, Regine. "Female Labour in the Japanese Coal-Mining Industry." In *Japanese Women Working*, edited by Janet Hunter, 98–121. London: Routledge, 1993.

———. "The Nation at Work: Gendered Working Patterns in the Taishō and Shōwa Periods." In *Gender, Nation and State in Modern Japan*, edited by Andrea Germer, Vera Mackie and Ulrike Wöhr, 141–162. New York: Routledge, 2014.

Michiko, Mae. "The Nexus of Nation Culture and Gender in Modern Japan: The Resistance of Kanno Sugako and Kaneko Fumiko." Translated by Leonie Stickland. In *Gender, Nation and State in Modern Japan*, edited by Andrea Germer, Vera Mackie and Ulrike Wöhr, 68–84. New York: Routledge, 2014.

Miyamoto, Takashi. "Excavating Memories Through HTML: The Internet and the Personal Archive of Maekawa Toshiyuki." *Waseda RILAS Journal*, no. 6 (October 2018): 553–559.

———. "Convict Labour and Its Commemoration: The Mitsui Miike Coal Mine Experience." *The Asia-Pacific Journal: Japan Focus* 15, no. 1 (January 2017): 1–14.

Molony, Barbara and Kathleen Uno. "Introduction." In *Gendering Modern Japanese History*, edited by Barbara Molony and Kathleen Uno, 1–35. Cambridge, MA: Harvard University Asia Center, 2005.

Nishinarita, Yutaka. "Introduction: Types of Female Labour and Changes in the Workforce, 1890–1945." In *Technology Change and Female Labour in Japan*, edited by Masanori Nakamura, 1–24. Tokyo: United Nations University Press, 1994.

———. "The Coal-mining Industry." In *Technology Change and Female Labour in Japan*, edited by Masanori Nakamura, 59–96. Tokyo: United Nations University Press, 1994.

Nite, Dhiraj Kumar. "Archives, Being and Representation: Studies into the Formation and Usage of Grassroots Archives." *Waseda RILAS Journal*, no. 6 (October 2018): 575–577.

Sand, Jordan. *House and Home in Modern Japan: Architecture, Domestic Space and Bourgeois Culture, 1880–1930*. Cambridge, MA: Harvard University Asia Center, 2003.

Schieder, Chelsea Szendi. "From Coal Miner's Wife to Historical Actor: The Personal Archive of Matsuo Keiko." *Waseda RILAS Journal*, no. 6 (October 2018): 545–552.

Schieder, Chelsea Szendi and Takashi Miyamoto. "Mining Grass-Roots Archives: The Japanese Experience." *Waseda RILAS Journal*, no. 6 (October 2018): 540–544.

Smith, W. Donald. "Digging Through Layers of Class, Gender and Ethnicity: Korean Women Miners in Prewar Japan." In *Women Miners in Developing Countries: Pit Women and Others*, edited by Kuntala Lahiri-Dutt and Martha Macintyre, 111–130. New York: Routledge, 2006.

———. "Sorting Coal and Pickling Cabbage: Korean Women in the Japanese Mining Industry." In *Gendering Modern Japanese History*, edited by Barbara Molony and Kathleen Uno, 393–422. Cambridge, MA: Harvard University Asia Center, 2005.

Sone, Sachiko. "Coal Mining Women Speak Out: Economic Change and Women Miners of Chikuho, Japan." In *Gender, Nation and State in Modern Japan*, edited by Andrea Germer, Vera Mackie and Ulrike Wöhr, 153–170. New York: Routledge, 2014.

———. "Japanese Coal Mining: Women Discovered." In *Women Miners in Developing Countries: Pit Women and Others*, edited by Kuntala Lahiri-Dutt and Martha Macintyre, 51–72. New York: Routledge, 2006.

———. "The Reversible World of Japanese Coalmining Women." *The Australian Journal of Anthropology* 18, no. 2 (August 2007): 207–222. https://doi.org/10.1111/j.1835-9310.2007.tb00089.x.

Yamamoto, Sakubei. "*Itcho Kiriha* [Single Coalface] June 1966." Translated by Nathan Johndro. Accessed October 24, 2018. http://www.y-sakubei.com/world_appl_e/gallery.html?mode=view&pnum=13&cat=13.

———. "*Meiji Chuki Asa no Kofu* [Miners Having Breakfast in the Mid-Meiji Era (1868–1912)] October 1967." Translated by Nathan Johndro. Accessed October 24, 2018. http://www.y-sakubei.com/world_appl_e/gallery.html?mode=view&pnum=1&cat=11.

———. "*Meiji, Taisho, Showa Ki-kozumi (Fufu de no Sagyo)* [Wooden Cribbing in the Meiji (1868–1912), Taisho (1912–1926), and Showa (1926–1989) Eras (Work by a Husband and Wife)] 1964–1967." Translated by Nathan Johndro. Accessed October 24, 2018. http://www.y-sakubei.com/world_appl_e/gallery.html?mode=view&pnum=11&cat=5.

———. "*Mukashi no Yamabito (Yufu no Kiridashi)* [People at Coal Pits (Yama) in the Old Days (Brave Woman Miner Cutting Coal Alone)] 1964–1967." Translated by Nathan Johndro. Accessed October 24, 2018. http://www.y-sakubei.com/world_appl_e/gallery.html?mode=view&pnum=5&cat=13.

———. "*Mukashi no Yama no Onna #14 (Tanju no Seikatsu; Fufu no Shokuji)* [Women at Coal Pits (Yama) in the Old Days #17 (Life in Row Houses; Meals of a Husband and Wife)] 1958–1963." Translated by Nathan Johndro. Accessed October 24, 2018. http://www.y-sakubei.com/world_appl_e/gallery.html?mode=view&pnum=143&cat=0.

———. "*Mukashi Yama no Onna #12 (Kantera o Sagete Nyuko Suru Boshi)* [Women at Coal Pits (Yama) in the Old Days #12 (Mother and Children Entering the Pit with Lamps in Hands)] 1958–1963." Translated by Nathan Johndro. Accessed October 24, 2018. http://www.y-sakubei.com/world_appl_e/gallery.html?mode=view&pnum=134&cat=0.

———. "*Mukashi Yama no Onna #17: Makitate de Kujibiki (Hako-tori no Kujibiki; Onna no Kamigata)* [Women at Coal Pits in the Old Days #17: Lottery Draw at the Landing (Lottery Draw for Empty Mine Cars; Women's Hair Style)] 1958–1963." Translated by Nathan Johndro. Accessed October 24, 2018. http://www.y-sakubei.com/world_appl_e/gallery.html?mode=view&pnum=80&cat=0.

Yoshida, Kayoko and Reiko Miyauchi. "Invisible Labour: A Comparative Oral History of Women in Coal Mining Communities of Hokkaido, Japan, and Montana, USA, 1890–1940." In *Mining Women: Gender in the Development of a Global Industry, 1670 to 2005*, edited by Jaclyn J. Gier and Laurie Mercier, 1–10. New York: Palgrave Macmillan, 2006.

CHAPTER 8

Conclusion

While there are concerted efforts to limit the impact of coal use on the environment, the future of coal fuels is here to stay, albeit in a transitional phase that will see the introduction of more fuels. Godo's introductory chapter is instructive on how coal fuelled industrialization and modernization, outperforming traditional fuels like timber. Coal has experienced at least two cycles of changes in Japan with the transition to oil in the 1960s and the decline of the domestic coal industry in Japan—the impact of which was studied in great detail in Shimazaki's chapter. While Shimazaki examined the highly important issue of re-employment and rehabilitation of laid-off workers, Tai Wei Lim looks at the physical impact itself on the community. Lim analyses the process and extent of gentrification in the major coal mining towns using observation studies and historical research related to their closures.

The second tidal change occurred with the rise of renewable and nuclear energies which became alternatives to coal. Since the 1960s, environmental awareness has strengthened considerably in Japan, resulting in societal pressures for the use of cleaner fuels. Coal was not a choice fuel in such a context. Against the backdrop of climate change, the transition away from coal use is even more acute, especially for the developed economies and the fast-developing ones. While phasing out coal is a long-term prospect, it cannot be done easily as coal is a cheap and plentiful fuel in Northeast Asia. Therefore, it cannot be completely eliminated from the energy mix. Coal's percentage of the energy mix may decline in Japan and other

T. W. Lim et al., *Coal Mining Communities and Gentrification in Japan*, https://doi.org/10.1007/978-981-13-7220-9_8

advanced economies, but its absolute amount is still expected to hold the ground or even increase in usage amount. Chapter 5 studied this aspect of the Northeast Asian coal industry in detail.

Godo's chapter provides a good context and foundation for studying the remains of Japan's early heavy industrial age, at a time of the emerging importance of iron, shipbuilding, and coal industries. His research proved to be helpful for a deeper understanding of the simultaneous industrial and agricultural activities visible when the research team for this book project visited the relics of Japan's industrialization and spotted how the mines co-existed with neighbouring agricultural communities in the east Hokkaido setting. Godo's analysis has comparative value for the agricultural, coal mining, and heavy industries, but his account also informs the readers of universal features found in all three industries, such as rationalization for greater productivity (e.g. the transition from "peasant time" to "inorganic time" of the factory system).

Correlating with Godo's analysis, the United Nations Education, Scientific and Cultural Organization (UNESCO) also divides Japanese coal industrial history into a number of historical phases, beginning with the pre-Meiji Bakumatsu *sakoku* isolation period in the last days of the Tokugawa government period in the 1850s and early 1860s where there was a hasty rush towards experimental trials in constructing iron and shipbuilding industries for military and security purposes.[1] Based mainly on experimental and textually based theoretical work that tries to copy and emulate Western manuals, these efforts mostly failed, accelerating the onset of the Meiji reforms when technology was directly imported and then indigenized to fit local conditions and to train local expertise. This was a major feature in Godo's chapter.

Godo's narrative is exemplified by the physical heritage sites that this volume's research team visited in Hokkaido. These sites still indicate a high level of integrity and authenticity in terms of conservation value while some are left in natural decline and ruin without tampering with the original structures. These sites represent the material link to Tai Wei Lim's chapter (Chap. 4) which detailed how the physical assets are utilized by local communities and authorities for tourism. The narrative of heavy industrialization becomes one of gentrification and memory archives. The main legal

[1] United Nations Education, Scientific and Culture Organization (UNESCO), "Sites of Japan's Meiji Industrial Revolution: Iron and Steel, Shipbuilding and Coal Mining" in the UNESCO website [downloaded on 1 Jan 2017], available at http://whc.unesco.org/en/list/1484

instrument used by the Japanese authorities to preserve such artefacts is the Law for the Protection of Cultural Properties for "regulating any development and change of the existing state of a designated place and under this law permission must be granted by the national government … [and] under the Landscape Act permission must be sought to change any Structure of Landscape Importance and owners of such structures must conserve and manage them appropriately."[2] Heritage and landscape preserved under this Act and other conservation guidelines are physical reminders of the past while museums and archives presented the human narratives behind the manmade structures.

Godo's attempt to highlight the human story in Japan's coal industrialization is significant not only for readers and visitors to the industrial relics but also to the international audience interested in Japan's modernization story. This resonates with conservation efforts by the Government of Japan. Kengo Iwamoto, Counsellor of the Cabinet Secretariat of Japan from the Department of Industrial Heritage World Heritage Inscription and a prominent individual advocating the preservation of the Meiji industrial sites, shares information about the sites to the world: "To properly communicate the value of these historic sites, it's vital to provide more than just visual information. We also must promote understanding of the culture, history and heart of the people of Japan that constructed them."[3] Japan's plan for the future are value-added for heritage lovers, history buffs, and the general public; Iwamoto revealed initiatives to design an upgraded website for international tourists and a 3D app that recreates areas that are out of bounds to the public, allowing them to enjoy views that are hitherto restricted.[4] Such future plans are likely to have a significant impact on the gentrification and development of local tourism detailed in Chap. 4 by Tai Wei Lim.

[2] United Nations Education, Scientific and Culture Organization (UNESCO), "Sites of Japan's Meiji Industrial Revolution: Iron and Steel, Shipbuilding and Coal Mining" in the UNESCO website [downloaded on 1 Jan 2017], available at http://whc.unesco.org/en/list/1484

[3] Kawasaki, Tami, "Japanese World Heritage An Unprecedented Miracle of Industrialization" in the Public Relations Office Government of Japan website [downloaded on 1 Jan 2017], available at http://www.gov-online.go.jp/eng/publicity/book/hlj/html/201509/201509_03_en.html

[4] Kawasaki, Tami, "Japanese World Heritage An Unprecedented Miracle of Industrialization" in the Public Relations Office Government of Japan website [downloaded on 1 Jan 2017], available at http://www.gov-online.go.jp/eng/publicity/book/hlj/html/201509/201509_03_en.html

Aside from relics and structures, Shimazaki's chapter is of utmost importance in studying the transitional phase. If Godo's chapter detailed the trials and tribulations of human achievements in developing agriculture and industries from a pre-modern setting as well as the human will to shape and craft nature to his/her advantage, Shimazaki's chapter details a different story, one that is both emotive (with the retrenchment and displacing of coal workers in Japan) and simultaneously cognitively reflective of the constant rationalization that goes on within the Japanese economy and industry. If Godo's chapter detailed the take-off stage of heavy industrialization (akin to Walt Whitman Rostow's take-off stage in his theory of the Five Stages of Economic Growth), then Shimazaki's chapter detailed its decline and rationalization in the coal industry. Both stories are equally important and significant in the narrating the overall story of the coal industry. Shimazaki's chapter has both sociological and policy implications. While the hard infrastructure detailed in Godo's chapter is constructed through human efforts and policy initiatives, Shimazaki's chapter detailed policies contributing to the rehabilitation of coal miners displaced by broader macroeconomic changes in energy use.

Sociological implications arise from the resettlement, rehabilitation, and reintegration of the families of the coal miners, entire communities, and even towns/villages/cities in the post-closure context. Rationalization has taken away the functional significance of the coal mining areas due to the emergence of the oil industry and foreign coal imports, but human communities bear the consequences in the wake of the massive economic changes brought about by closures. Shimazaki detailed the displacement of such communities and the measures taken to assist with reintegration back into the general economy. Policy instruments provided the motivation for coal miners to cope with changes.

Shimazaki's works have also detailed the sort of internal solidarity enjoyed by these communities through personal ties, collegiality, union camaraderie, and common experiences socialized through the course of coal mining work. Such personal ties are enduring and highly visible throughout the Hokkaido visits made by the research team for this publication. While the former coal miners and their communities are now scattered all over Japan or absorbed into neighbouring cities/towns, their personal ties remain strong through word of mouth about each other's well-being, get-togethers, and reunions. Shimazaki terms these personal ties as "informal support" supplementing rehabilitation efforts that policies alone find difficult to fulfil.

Whether in studying human relations or heritage structures, Hokkaido is an important setting for studying the personal ties and aftermath of mine closures as Shimazaki noted that coal industrial development in Japan was:

> Initially focused in the Kyushu area, it was later expanded to include Hokkaido. As a result, decline in the coalfield regions began in Kyushu. In Hokkaido, new coal mines were still being established up until the 1980s.[5]

In other words, Hokkaido was a late developer in coal industrialization and the memories and oral accounts of its mine closures can still be recalled by former miners. The research team for this volume met a number of such miners. Hokkaido's coal mining heritage sites and relics are also more recent and it still has one operational coal mine in Kushiro. There are features diachronic with coal mines built in earlier time periods, like those found in Kyushu, while other characteristics are synchronic and demonstrate similarities with other Japanese coal mining regions.

Most importantly, for Shimazaki's study, the coal miners' personal networks are self-sustaining at the current moment. The coverage of human relations and networking is a commonality that binds both Godo and Shimazaki's analyses and underlines the sociological elements in their chapters. For example, Godo detailed the massive changes in human relationships between landlords and tenants, the urbanization of wealthy absentee landowners, socialization of modern ideas through the public school education system for adapting to a factory system, and so on. These massive social changes are akin to Shimazaki's account of post-closure rehabilitation of miners, forging strong bonds of solidarity out of a common socialization experience that was equally traumatic but also positively transformative for many. These sociological elements bind Godo and Shimazaki's chapter chronologically by detailing the rise and decline of the coal mining industry. Institutionally, the advocate groups of *Nokyo* (agricultural cooperatives) in Godo's analysis and the mining unions in Shimazaki's chapter also resonate with each other as institutions that collectively aggregate interests of their individual constituents.

[5] Shimazaki, Naoko, "Support for Workers Displaced in the Decline of the Japanese Coal Industry: Formal and Informal Support" dated Spring 2015 in Japan Labor Review Vol. 12 No. 2 (Japan: Japan Labor Review), p. 11.

Tai Wei Lim's chapter on Northeast Asia (Chap. 5) spotted common trends in the Northeast Asian coal mining industry. Lim highlighted how the Great East Japan Earthquake and the Fukushima nuclear leakage aftermath in 2011 have seen the country importing more coal. In fact, Japan purchased more liquefied natural gas (LNG) and coal from Australia with an aim of diversifying fossil fuel sources without any overdependence (Japan imports 95% of its energy needs). In this way, coal is important as an affordable component of energy diversification.[6] Indicative of coal's persistent relevance, Japan, China, and South Korea are financing coal-powered electricity-generating stations, accounting for 18 such investments in Indonesia since January 2010, providing funding up to US $16.7 billion.[7] The Export-Import Bank of China (China Exim) participated in seven such coal financing scheme, while the Japan Bank for International Cooperation (JBIC) and the China Development Bank (CDB) are involved in five projects each; the Korea Development Bank (KDB) is involved in one and the Export-Import Bank of Korea (Korea Eximbank) is participating in one project alongside JBIC.[8] These trends indicate that coal use is not about to disappear in Northeast Asia but their extraction has been outsourced to other regional countries in Southeast Asia, Oceania, and other coal-rich regions in the world.

Closing Remarks

Weaved into all the chapters, there are human stories of the Japanese coal mining industry. In Godo's chapter, the human-centred historical narrative is that of the individual peasant and his collective making a transition from peasantry to modern industrial workers. Perhaps the symbolic human-centred transformation lies in the transformation of the peasants'

[6] McHugh, Babs, "Japanese government planning to build 45 new coal fired power stations to diversify supply" dated 1 February 2017 in the ABC News website [downloaded on 1 Feb 2017], available at http://www.abc.net.au/news/rural/2017-01-31/japan-coal-power-plants/8224302

[7] Win, Thin Lei, "Japan, China, South Korea bankroll Indonesia's coal despite Paris pledge" dated 19 July 2017 in the Reuters website [downloaded on 19 July 2017], available at http://www.reuters.com/article/us-china-pollution-smog/chinas-hebei-vows-coal-curbs-ahead-of-winter-smog-war-idUSKCN1BQ0C1

[8] Win, Thin Lei, "Japan, China, South Korea bankroll Indonesia's coal despite Paris pledge" dated 19 July 2017 in the Reuters website [downloaded on 19 July 2017], available at http://www.reuters.com/article/us-china-pollution-smog/chinas-hebei-vows-coal-curbs-ahead-of-winter-smog-war-idUSKCN1BQ0C1

concept of time to a rationalized production schedule. Shimazaki's chapter is even more emotive, with individual, family groups, and entire towns displaced by macroeconomic changes. The individual and collective trials and tribulations of coal miners are highlighted in this story of economic transition of an entire industry. Tai Wei Lim's chapter lays out the environmental changes in the landscape and how it transforms human experiences in appreciating aesthetics and ethical/moral rejuvenation with a cleaner, greener, and more creatively artistic environment in gentrified spaces. Finally, the last two chapters before conclusion are perhaps the most human and emotive story of all the chapters. Chapter 6 details the expression of human emotions in the coal mining industry through poetry, films, and literary devices.

In the final chapter (Chap. 7) before the conclusion, Yiru Lim's gendered analyses of the mining industry questioned and critiqued existing modes of understanding. Lim tackled the histories and stories of coal mining women in Japan which have, for the most part, been a neglected area of study. Previously, most historical surveys or sociological studies of Japan's economic miracle, technological progress, and industrial growth and development have looked at the coal mining industry as almost synonymous with the male miner, choosing to discuss the coal industry from such a perspective, presuming, inaccurately, that gendered experiences can be universalized. This chapter utilized a gendered lens to understand the socio-cultural and historical contexts and working conditions under which Japanese women miners laboured, and, consequently, to analyse points of resistance that women displayed within this environment. Lim's chapter concentrated on social history and community relations and dynamics to understand gender relations. The counter-narratives in her chapter represented history by focusing on a specifically working-class narrative and its "gender dimension" which is set up as a contrast to the post-war image of Japan as a nation of the middle class.

Together, these chapters weave coherently into a human-centred story of the trials and tribulations of an industry, the human agencies and actors within it, and the transformation of disused spaces for the renewal and reinvigoration of the human spirit and the creative expressions of their sense of aesthetics and beauty. In another way, there is a cyclical pattern that binds together all chapters by the four authors in the publication. Godo detailed the evolution of pre-modern agricultural and primary products industry to heavy industrialization powered by coal. Shimazaki's chapter portrays an obsolete coal mining landscape that is primed for closure with coal miners as Japan enters a post-industrial phase of development with concerns about the environment. But Tai Wei Lim's chapter details some former mining regions' reversion back to agricultural activities, this time in high value-added ventures like wineries and cultivating premium branded products like Yubari melons. The narrative represents a cyclical portrayal of a tussle between Man and Nature in crafting the use of terrestrial spaces that both of them share. As some coal mining towns depopulate or age, nature reclaims those spaces by growing over rusting and abandoned infrastructures. Bears, for example, now roam the abandoned Urahoro Coal Mine, a fact apparent to this publication's research team when they visited the area. Nature retakes the initiative in the contestation of land use and the beneficiaries are some local communities who can enjoy a cleaner environment, sustainable use of natural resources, and a diversity of wildlife, evident in locations like Kushiro River.

Bibliography

Kawasaki, Tami, "Japanese World Heritage: An Unprecedented Miracle of Industrialization" in the Public Relations Office Government of Japan website [downloaded on 1 Jan 2017], available at http://www.gov-online.go.jp/eng/publicity/book/hlj/html/201509/201509_03_en.html

McHugh, Babs, "Japanese government planning to build 45 new coal fired power stations to diversify Supply" dated 1 February 2017 in the ABC News website [downloaded on 1 Feb 2017], available at http://www.abc.net.au/news/rural/2017-01-31/japan-coal-power-plants/8224302

Shimazaki, Naoko, "Support for Workers Displaced in the Decline of the Japanese Coal Industry: Formal and Informal Support" dated Spring 2015 in Japan Labor Review Vol. 12 No. 2 (Japan: Japan Labor Review), pp. 6–27.

United Nations Education, Scientific and Culture Organization (UNESCO), "Sites of Japan's Meiji Industrial Revolution: Iron and Steel, Shipbuilding and Coal Mining" in the UNESCO website [downloaded on 1 Jan 2017], available at http://whc.unesco.org/en/list/1484

Win, Thin Lei, "Japan, China, South Korea bankroll Indonesia's coal despite Paris pledge" dated 19 July 2017 in the Reuters website [downloaded on 19 July 2017], available at http://www.reuters.com/article/us-china-pollution-smog/chinas-hebei-vows-coal-curbs-ahead-of-winter-smog-war-idUSKCN1BQ0C1

Index[1]

[1]Note: Page numbers followed by 'n' refer to notes.

T. W. Lim et al., *Coal Mining Communities and Gentrification in Japan*, https://doi.org/10.1007/978-981-13-7220-9

The manufacturer's authorised representative in the EU is Springer Nature Customer Service Centre GmbH, Europaplatz 3, 69115 Heidelberg, Germany. If you have any concerns regarding our products, please contact ProductSafety@springernature.com

Printed and bound by CPI Group (UK) Ltd, Croydon, CR0 4YY
25/09/2026
02222897-0008